Ethical Password Cracking

Decode passwords using John the Ripper, hashcat, and advanced methods for password breaking

James Leyte-Vidal

Ethical Password Cracking

Group Product Manager: Pavan Ramchandani

Publishing Product Manager: Prachi Sawant

Book Project Manager: Ashwin Kharwa

Senior Editor: Isha Singh, Sayali Pingale

Technical Editor: Nithik Cheruvakodan

Copy Editor: Safis Editing

Indexer: Subalakshmi Govindhan

Production Designer: Prafulla Nikalje

DevRel Marketing Coordinator: Marylou De Mello

First published: July 2024

Production reference: 1120624

Published by Packt Publishing Ltd.

Grosvenor House

11 St Paul's Square

Birmingham

B3 1RB, UK

ISBN 978-1-80461-126-5

www.packtpub.com

To my incredibly patient family, as we weathered personal and professional challenges to deliver this book (and a real hurricane as well!). Rachel, Abigail, Brandon, Catherine, and Ethan – I love you all. This work is as much yours as it is mine.

And also, to the incredibly patient team at Packt Publishing, thank you for sticking with me through all this.

Contributors

About the author

James Leyte-Vidal is a 20-plus-year veteran of the computer security industry. After a self-taught career in IT, James worked on a computer security incident that changed his career trajectory to security.

James consults independently and has worked for Fortune 100 companies in various roles, including security architecture, penetration testing, compliance, policy, and much more.

James is also an instructor at the SANS Institute, a global provider of information security training, and a co-author of three SANS courses: *SEC467: Social Engineering for Security Professionals*, *SEC556: IoT Penetration Testing*, and *SEC617: Wireless Penetration Testing and Ethical Hacking*.

When not actively doing security work, James can often be found tinkering with hardware or spending time with his family.

About the reviewers

Matt Edmondson is a principal SANS instructor and author of the SEC497 Practical OSINT course. He has over 21 years of experience as a federal agent and is the founder of Argelius Labs, where he has experience in helping numerous organizations large and small understand their external attack surface and monitor for emerging threats. He is a multiple-time speaker at Black Hat and has featured in multiple publications, including Wired and the official Raspberry Pi magazine.

Rich Robertson has held several leadership and individual contributor roles throughout his 20+ years in IT and cybersecurity. His career covers a variety of industry verticals, including finance, entertainment, retail, and technology. Rich's passions include mobile devices, hardware hacking, data mining, and uncovering logic flaws. He has worked to secure the technology that powers everything from warehouses to cruise ships to dolphin tanks. Rich received a BSc in business administration and an MBA from Webber International University, and recently his MSc in FinTech at the University of Central Florida. He also holds multiple industry certifications, such as CISSP, ITPM, and SANS GCIH and GPEN.

I would like to thank James for the opportunity to review his book. Thank you also to my beautiful wife, Stacey, for the love, support, and patience throughout the craziness, and my son, Tyler, for helping me pull this off. To my parents, thank you for your faith in me and letting me build my first computer, which jumpstarted my whole career in tech. I would also like to thank the rest of my family, friends, colleagues, and more (JM & LM, LJ, and Mika H) who have pushed me to become who I am today.

Table of Contents

Part 2: Collection and Cracking

Part 3: Conclusion

11

Preface

In this book, we will introduce you to the concepts behind password cracking, as well as make you familiar with the common tools used for this work. After that, we will examine common technologies where these tools may be needed, and show you how to retrieve and crack hashes for those particular technologies. While not exhaustive, these examples of applications of the tools will prepare you to deal with other types of password hashes you may work with later on.

Who this book is for

This book is designed for those with an interest in password cracking, but you do not necessarily need any experience.

What this book covers

Chapter 1, *Password Storage: Math, Probability, and Complexity*, provides an introduction to the concepts behind password cracking.

Chapter 2, *Why Crack When OSINT Will Do?*, provides a treatment on **Open Source Intelligence (OSINT)** as an alternative to password cracking in some cases.

Chapter 3, *Setting Up Your Password Cracking Environment*, provides an introduction to the tools needed for password cracking.

Chapter 4, *John and Hashcat Rules*, provides an introduction to permutation rules in John and hashcat and how they work.

Chapter 5, *Windows and macOS Password Cracking*, covers obtaining, formatting, and cracking Windows and macOS password hashes.

Chapter 6, *Linux Password Cracking*, covers obtaining, formatting, and cracking Linux password hashes.

Chapter 7, *WPA/WPA2 Wireless Password Cracking*, covers obtaining, formatting, and cracking hashes for WPA and WPA2 Wi Fi networks.

Chapter 8, *WordPress, Drupal, and Webmin Password Cracking*, covers obtaining, formatting, and cracking hashes for the WordPress, Drupal, and Webmin platforms.

Chapter 9, *Password Vault Cracking*, covers obtaining, formatting, and cracking KeePass, LastPass, and 1Password vault passwords.

Chapter 10, *Cryptocurrency Wallet Passphrase Cracking*, covers obtaining, formatting, and cracking Bitcoin, Litecoin, and Ethereum wallet passwords.

Chapter 11, *Protections against Password Cracking Attacks*, discusses potential defenses against password cracking.

To get the most out of this book

You will need a working knowledge of Windows and Linux command-line syntax and common non-OS tools, such as Git.

Software/hardware covered in the book	Operating system requirements
hashcat	Windows, macOS, or Linux
John the Ripper	Windows, macOS, or Linux

> **Disclaimer**
> The information within this book is intended to be used only in an ethical manner. Do not use any information from the book if you do not have written permission from the owner of the equipment. If you perform illegal actions, you are likely to be arrested and prosecuted to the full extent of the law. Packt Publishing does not take any responsibility if you misuse any of the information contained within the book. The information herein must only be used while testing environments with proper written authorizations from the appropriate persons responsible.

Conventions used

There are a number of text conventions used throughout this book.

`Code in text`: Indicates code words in text, database table names, folder names, filenames, file extensions, pathnames, dummy URLs, user input, and Twitter handles. Here is an example: "Once we have access to `wallet.dat`, we can use this file to extract the passphrase hash for cracking."

Any command-line input or output is written as follows:

```
python3 /opt/john/run/bitcoin2john.py wallet.dat
```

Bold: Indicates a new term, an important word, or words that you see onscreen. For instance, words in menus or dialog boxes appear in **bold**. Here is an example: "Drilling into the **LastPassData** table by selecting **Browse Table**, we see several rows of interest to us."

> **Tips or important notes**
> Appear like this.

Get in touch

Feedback from our readers is always welcome.

General feedback: If you have questions about any aspect of this book, email us at `customercare@packtpub.com` and mention the book title in the subject of your message.

Errata: Although we have taken every care to ensure the accuracy of our content, mistakes do happen. If you have found a mistake in this book, we would be grateful if you would report this to us. Please visit `www.packtpub.com/support/errata` and fill in the form.

Piracy: If you come across any illegal copies of our works in any form on the internet, we would be grateful if you would provide us with the location address or website name. Please contact us at `copyright@packt.com` with a link to the material.

If you are interested in becoming an author: If there is a topic that you have expertise in and you are interested in either writing or contributing to a book, please visit `authors.packtpub.com`.

Share Your Thoughts

Once you've read *Ethical Password Cracking*, we'd love to hear your thoughts! Scan the QR code below to go straight to the Amazon review page for this book and share your feedback.

https://packt.link/r/1804611263

Your review is important to us and the tech community and will help us make sure we're delivering excellent quality content.

Download a free PDF copy of this book

Thanks for purchasing this book!

Do you like to read on the go but are unable to carry your print books everywhere?

Is your e-book purchase not compatible with the device of your choice?

Don't worry!, Now with every Packt book, you get a DRM-free PDF version of that book at no cost.

Read anywhere, any place, on any device. Search, copy, and paste code from your favorite technical books directly into your application.

The perks don't stop there, you can get exclusive access to discounts, newsletters, and great free content in your inbox daily

Follow these simple steps to get the benefits:

1. Scan the QR code or visit the following link:

https://packt.link/free-ebook/9781804611265

2. Submit your proof of purchase.
3. That's it! We'll send your free PDF and other benefits to your email directly.

Part 1: Introduction and Setup

In this part, we will introduce you to the concepts behind password cracking, alternative means to achieve our goals such as OSINT, and how to set up and configure your cracking environment.

This part has the following chapters:

- *Chapter 1, Password Storage: Math, Probability, and Complexity*
- *Chapter 2, Why Crack When OSINT Will Do?*
- *Chapter 3, Setting Up Your Password Cracking Environment*
- *Chapter 4, John and Hashcat Rules*

Password Storage: Math, Probability, and Complexity

Password cracking has become a storied element of information security testing, from the days of utilities such as Cain and Abel to more modern tools such as hashcat. While the tools and techniques have changed over the years, the principles behind password cracking remain largely unchanged.

Password cracking can involve many use cases, from recovering access to a system after the user has left a company to penetration testing and red team use cases, where we use password cracking to prove (or disprove) the security of our access control mechanisms.

In this chapter, we're going to cover the following main topics:

- What is password cracking?
- How are passwords stored and used?
- Why are some passwords easier to crack than others?

What is password cracking?

Password cracking is the process of recovering a secret from scrambled (typically encrypted or hashed) text. This very broad term encompasses many types of password storage and scrambling. As such, not all password-cracking operations are created equal – some passwords, as well as methods of password storage, are easier to crack than others. We will discuss this more throughout this book.

Password cracking can be broken down into various approaches to attempt to recover the secret:

- Dictionary-based
- Combination

- Brute force
- Hybrid
- Partial knowledge, also known as mask attacks

Let's discuss each of these in turn.

Dictionary-based attacks

Dictionary-based attacks, as you might have guessed based on the name, use a list of words or phrases as password candidates – the potential password we will test to see if it is the correct password. This list is informally referred to as a *dictionary*, even though it may or may not contain dictionary words. The wordlist may not resemble a dictionary much at all. This term is mostly a holdover to earlier times when many passwords were based on dictionary words, before password complexity requirements (such as adding uppercase letters, numbers, and symbols to a password) were common.

Speaking of complexity requirements, it seems like traditional dictionary words would not be as effective as password candidates during a password-cracking operation due to complexity requirements becoming more commonplace. We'll address that in the upcoming sections.

Constructing a wordlist for a dictionary attack can be simple or a time-consuming effort. However, in many cases, spending time upfront for a good wordlist tailored to your target may reap dividends at cracking time. The tradeoff here is that your wordlist may not be as reusable for other password-cracking situations. We'll discuss using **open source intelligence** (**OSINT**) to help build a wordlist in *Chapter 2, Why Crack When OSINT Will Do?*

A good and fairly large wordlist to start with is often the **RockYou wordlist**. This is named after the breach of the RockYou company in 2009, where over 32 million user credentials were exposed. While available in several places on the internet, a common location to download the RockYou wordlist is `https://github.com/brannondorsey/naive-hashcat/releases/download/data/rockyou.txt`. This list contains over 14 million unique password candidates and is also included in many common penetration testing distributions, such as Kali Linux (available at `https://www.kali.org/get-kali/#kali-platforms`) and Slingshot Linux (available at `https://www.sans.org/tools/slingshot/`).

Combination attacks

Combination attacks take two wordlists as input and concatenate (append together) one password candidate from each list to create the password candidate for testing purposes. For example, one wordlist might contain the words `word1` and `word2`, while the second wordlist might contain the words `word3` and `word4`.

In this scenario, a combination attack would use a word from both lists to create potential password candidates, such as `word1word3`, `word1word4`, `word2word3`, and `word2word4`.

Current guidance from the **National Institute of Standards and Technology** (**NIST**) recommends password length over complexity for the best resistance to password cracking. This helps encourage our users to create a password that is easy to remember but hard to crack and reflects the current guidance from NIST. This can be performed by stringing several dictionary words together and adding a mnemonic to help the user remember the password. This is only one approach, but this example points out – in conjunction with the current NIST guidance – that combination approaches to password cracking may be more effective as more users follow the guidance to shift to passphrases.

That being said, some standards may slow the adoption of longer, less complex passphrases. For example, the **Payment Card Industry Data Security Standard** (**PCI-DSS**) standard, which is required for merchants processing credit card data, requires 12-character passwords, as well as letters and numbers for passwords associated with accounts that have access to cardholder data.

Brute-force attacks

Brute-force attacks do exactly what their name suggests – every position in the password candidate is filled with every possible candidate for that position. For example, if a password can only be eight characters long, a brute-force approach might attempt `aaaaaaaa` as the password candidate, then attempt `aaaaaaab`, and so on, until the possibilities for the password are all attempted – *exhausted*. The problem with this approach is that once a password reaches any reasonable length, the time to perform this style of attack becomes untenable. Additionally, the number of character sets available to use for the password (lowercase, uppercase, numbers, and symbols) will also greatly increase the number of guesses to complete this kind of attack.

The good news for password cracking is that it is possible to crack any password with this approach. However, the amount of time it would take with today's computing power makes it essentially folly for larger passwords or more complex (more time-consuming for each password guess) algorithms.

Hybrid attacks

Hybrid attacks merge some of the characteristics of combination attacks and brute-force attacks. A hybrid attack uses a wordlist as its base, then modifies the words in the wordlist by adding one or more characters to the word and brute-forcing the character space associated with that. As an example, let's say I have the following word from my wordlist:

```
banana
```

However, I know the password policies of my target require a number in every password. I might try a hybrid attack that takes my word from my wordlist and adds a number after the word. So, now, my password candidates are as follows:

```
banana1
banana2
banana3
banana4
```

This allows us to test environments more effectively where users often append (add to the end) or prepend (add to the beginning) some base dictionary word for their password.

Partial knowledge, also known as mask attacks

Mask attacks leverage the idea that we partially understand the format used to construct a password to create a brute-force-like approach that is sped up due to assumptions we make about the password format.

An example will be helpful here. Let's say that we are testing passwords for a company that requires one uppercase character, one lowercase character, and a number for their passwords. This is a common password complexity requirement in many companies, and many users will meet this requirement by taking a word (dictionary or otherwise), capitalizing the first letter of the word, and appending one or two numbers to the word.

Incidentally, this type of password requirement, along with 90-day password rotation intervals, can lead to the dreaded *season-year* password, where users will set their password to the name of the current season (Spring, Summer, and so on) and append a two or four-digit year to the password (Spring22/ Spring2022, Summer22/Summer2022, and so on).

These complexity requirements may lead us to construct a mask for the password that assumes the user will choose a password that starts with a capital letter, then has five or six characters of lowercase letters, and ends with two or two digits from base10 numbering (0-9). This mask will attempt to brute-force any passwords meeting these lengths and criteria. While this will not retrieve every password in a given list, this approach historically yields high percentages of cracked passwords since this approach is a common one for users to take when constructing passwords.

> **Important note**
> We will suggest better methods for password construction and mitigations in *Chapter 11*.

How are passwords stored and used?

While it may seem simple, how a password is stored on a system can have a huge effect on its ability to be recovered via password-cracking operations and how long this can take.

> **You don't always need to crack!**
>
> Most passwords are stored in authentication systems via some process that renders the password difficult to recover. However, it is not unheard of to come across systems that do not protect user credentials appropriately. You may recall that earlier in this chapter, we discussed the RockYou breach. In the case of RockYou, the company stored user passwords in plaintext (no hashing or encryption), which made recovering user passwords trivial. This meant that once user passwords were made publicly available, they were completely exposed – no password cracking or other complex operations were required; they were simply there for the taking.

Let's talk about the two types of responsible password storage that we typically see: **hashing** and **encryption**.

Hashing

The idea behind password hashing is to store the user's password so that it cannot be retrieved by anyone. There are several advantages to this approach:

- For the company that stores the password, this represents a strong level of due diligence and *may* provide some protections legally

- Passwords cannot be reverted to plaintext (the original password) from hash values, which means malicious insiders with access to the password storage cannot retrieve the password

- The existence of standard functions to perform this hashing in many application frameworks means it is easy to implement

At its core, hashing takes a string of plaintext (the password) and converts it into a fixed-length string of unreadable data. This value cannot be reverted to plaintext, which is one of the core differences between hashing and encryption. Also, this hashing process will always return the same value for the same input; this is known as being **deterministic**. Some types of hashing can also add a **salt**, which adds additional entropy (randomness) to the generation of the hash value. This salt will be different for every password, which can negate the effectiveness of **precomputation attacks** – a type of attack that generates all possible hashes in advance of a cracking operation (you may have heard of rainbow tables, which are one type of precomputation attack). Hashing algorithms vary in terms of the number of rounds (hash operations) used to create the hash to be stored, the output length, and several other factors. We will discuss various hashing algorithms later when we dive into different types of password retrieval.

In the case of hashing, passwords are validated during the authentication process by taking the password from the user, hashing it, and comparing it against the stored hash. If they match, the password is correct; if they do not, the password that was entered was incorrect. Once again, hashing further protects the plaintext password during this process by ensuring the plaintext password is never handled by the system after hashing.

Encryption

Encryption differs from hashing in that the ciphertext (the product of the encryption algorithm) can be reversed back to the original plaintext (the password). To do this, one or more encryption keys must be generated and used for encrypt and decrypt operations.

Encryption has some liabilities for use as password storage. The most prominent one is that the ciphertext is reversible, which means that a malicious insider or an external party can retrieve the plaintext passwords if they can obtain the ciphertext and the encryption key(s). Additionally, because it is used in encrypt and decrypt operations, the key(s) must be retrievable, which further increases the potential for mishandling and/or disclosure of the keys.

> **Easy check for encryption as password hashing (or worse)**
>
> Have you ever forgotten a password and used a **Forgot Password** link or workflow in an application? Odds are, you probably have. If you have ever used the **Forgot Password** functionality and had your password sent to you via email or some other cleartext method (rather than being prompted to set a new password), this means that your password is stored on that system in an encrypted format. If password hashing was in use, they would not be able to retrieve your plaintext password. Well, there's one other possibility – the system is storing your password in cleartext, similar to what RockYou did. We have seen how that is a very bad idea, but unfortunately, it is sometimes done.

In the case of authentication with encrypted passwords, the ciphertext can be compared (similar to authentication with hashing in use), or the password can be decrypted and compared to validate the password provided by the user.

While encryption has been noted here for completeness, it is not at all optimal to use encryption for password storage and is not recommended in the NIST 800-53 standards.

Why are some passwords easier to crack than others?

There are several reasons for this, but it boils down to one thing: how long it takes for a system to guess the password correctly. If we can create passwords that increase the amount of time for this to happen, we are creating a password that is more difficult to crack. If we create passwords that decrease the amount of time needed for password guessing to be successful, we are creating a password that is easier to crack.

So, what are some of the factors that make a password easier (or more difficult) to crack? Some of the most important are as follows:

- Password length
- Password complexity
- Time to hash/encrypt the password

Let's talk about each of these in turn.

Password length

Password length is often thought about by the end user in terms of the bare minimum. In other words, if a system requires an eight-character password at a minimum, many users will select an eight-character password.

At the time of writing, NIST maintains its password recommendations at an eight-character minimum. This is noted in NIST **Special Publication (SP)** 800-53B and is updated from time to time. However, NIST also notes that systems should accept a password from a user of at least 64 characters.

Let's think about that eight-character password for a moment. How many guesses would I need to make to determine someone's eight-character password? The answer, as with so many things in information security, is that *it depends*. Let's start with a simple character set that consists of the 26 (lowercase) letters of the English alphabet. The number of guesses required to successfully determine this password is represented by x to the power of y or x^y, where x is the possible characters in each position of the password, and y is the number of total characters in the password. For our 26-character lowercase password, which is 8 characters in length, it will take 26^8 guesses, or 208,827,064,576 guesses. Note that this is the *maximum* number of guesses – this represents someone guessing every possible password and only being successful on the last guess. This is a lot of possible guesses! But does this mean this password is secure? Again, it depends. How quickly can we try a guess and validate if it is or is not the password? Even milliseconds less or more per guess can have a large impact on the overall time to work through all the possibilities.

What if we choose a password length that is longer than the minimum recommendation from NIST? What about nine characters with the same lowercase English alphabet? That's 26^9 or 5,429,503,678,976. This is, as you might expect, 26 times more guesses than what we needed to make for an eight-character password.

By the time we get to a 12-character password, with our same 26-character set, we are looking at 26^{12}, or 95,428,956,661,682,176 (also known as roughly 95 quadrillion guesses). This is 456,976 times the number of guesses required for an eight-character password!

Visualizing this in a graph (see *Figure 1.1*), we can see an exponential growth of guesses required for every character increase of the password length:

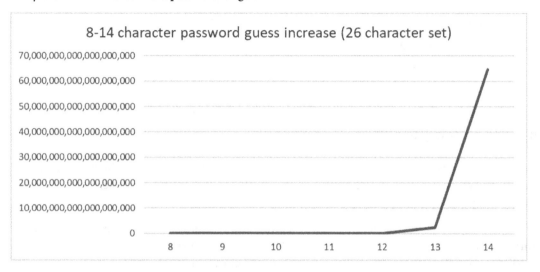

Figure 1.1 – Number of guesses for 8 to 14-character passwords (26 possible characters)

For those building secure systems, this is good, and this means every character counts when it comes to password length. The longer a password is, the longer it will take to crack, and the more secure (resistant to cracking) it is.

Password complexity

The idea behind password complexity, like password length, is to make a password more resistant to cracking. However, complexity takes a different approach – for every character in the password, we increase the possible characters that can be used to fill that spot. Let's see how this works in practice by revisiting our math in the previous section.

If we add uppercase English alphabet characters to our lowercase English alphabet characters, we get 52 possible characters. So, now, our 8-character password will require 52^8 guesses, or 53,459,728,531,456. Here, adding an additional 26 characters significantly increases the number of guesses. Furthermore, because this is an exponential operation, the increase in the number of guesses per character can be visualized in the same way as the 26-character password as length increases (see *Figure 1.2*):

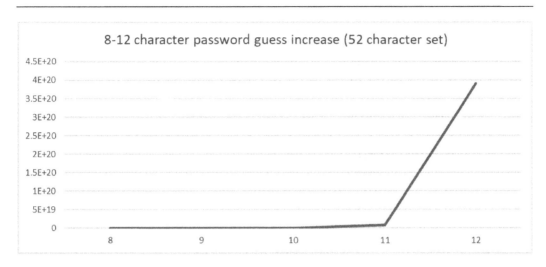

Figure 1.2 – Number of guesses for eight to 12-character passwords (52 possible characters)

As shown in *Figure 1.1*, increasing password length increases the total guesses required to identify a password. Likewise, in *Figure 1.2*, we can see that increasing the complexity of the password increases the guesses required, and increasing both length and complexity raises the number of guesses required even faster! So, which is better? Or should we use both? For this answer, we need to look at the math, and then follow it up with psychology.

An eight-character password with uppercase and lowercase letters requires 53,459,728,531,456 (53 trillion) maximum guesses. A 10-character lowercase-only password will require almost three times as many guesses – 141 trillion. Now, let's move on to the psychology. Which will be easier for a human to remember – an all-lowercase series of characters, or a series of uppercase and lowercase characters? One of two things will likely happen:

- The user will create a password that's easy for them to remember by capitalizing the first letter of the password and leaving the rest lowercase. This is trivial to address in cracking and subverts the point of adding the additional character set. If the first letter of the password is capitalized, there are 26 possible choices, meaning the same number of choices when we use lowercase characters. If the user then leaves the rest of the password lowercase, there are only 26 possible choices per character there as well. In this scenario, with an eight-character password, we have 26^8 possibilities instead of 52^8 possibilities – *the same number as if the password had just been lowercase to begin with*!

- The second possibility is that the user creates a hard-to-remember password and writes it down either on paper or in a password manager. While the use of a password manager is generally desired behavior, writing a password down where it might be discovered is not.

So, where does this leave us? The human mind will find an all-lowercase password to be easier to remember than a series of uppercase and lowercase letters, a series of upper and lowercase letters and numbers, or a series of upper and lowercase letters, numbers, and symbols. We can increase the length of all lowercase passwords and still create a password that is resistant to cracking. This is the current NIST recommendation – the current revision of SP 800-53B suggests that creating a password should *not* require password composition rules to be used (section 5.1.1.2).

Time to hash/encrypt the password

The third major factor in creating passwords that are resistant to cracking is not in the selection of the password itself, but rather the computational operations to create the hash and how long they take. Think about the number of guesses required for the various types of passwords we discussed earlier. Each of those guesses takes a non-zero amount of time to perform. We must calculate the hash for that password candidate, and then compare it against the known hash to see if they match (meaning our password candidate is our password).

If this operation takes a full second, instead of half a second, the overall time for the cracking process is doubled. In reality, guesses will occur much faster than that but for the sake of illustration, you can see how that makes a huge difference against the number of guesses we are dealing with in these scenarios.

Hashing algorithms are designed to be fast. Hashing is a common computational operation for comparisons, and we want them to be fast. However, we want password hashing specifically to be slow – we want it to be as slow as we can reasonably get away with. The slower the password hashing operation, the more resistant the implementation will be to cracking by making each attempt more computationally expensive. Password hashing algorithms such as PBKDF2 use common hashing algorithms such as SHA-512 but run many rounds of that hashing algorithm to increase the time to create the password hash.

While increased time per hash will result in a slower cracking operation, the cracker can attempt to offset this by increasing the number of hashes they perform per second, either by increasing the computational power used by the cracking process or distributing the load of the cracking operation across multiple computational engines. In *Part 2, Password Cracking Types and Approaches*, we will look at the overall speed of various cracking operations based on the types of hashes we are cracking.

A word on "ethical" password cracking

Regardless of the approach, the objective of this book is to help you with the tools and techniques you need to recover passwords, whether you are in a penetration test/red team operation, you are recovering a password of a user backup for someone who has unfortunately passed away, or anything in between.

The important caveat here is that this book focuses on ethical password cracking. The purpose of this book is not to help you circumvent laws or perform illegal activities. Rather, its purpose is to give you what you need to be successful in pointing out flaws in penetration testing engagements, or other approved means to enable your business.

Please ensure you consult with your legal counsel and/or company counsel before performing these techniques against passwords in your company.

Summary

In this chapter, we introduced you to the concept of password cracking, the various types of cracking attacks, how passwords are stored and used, and some of the reasoning behind what makes a stronger password. With this, you have laid the groundwork to get started with various types of password cracking.

However, wouldn't it be easier if we never had to crack a password at all? In some cases, we can, because of readily available information such as previous data breaches and poor password practices such as password reuse. In the next chapter, we will examine how to use OSINT to find information from previous breaches or to build custom wordlists for specific targets.

2

Why Crack When OSINT Will Do?

While the focus of this book is on password cracking – the recovery of plaintext credentials given their scrambled and unreadable (be it hashed, encrypted, or otherwise) content – we do face challenges along the way. In some cases, password-scrambling algorithms may be too time consuming to recover plain text from, and in others we may exhaust our possible guesses and still not discover a password.

In these situations, there may be an easier alternative – leveraging our known public resources to determine more about the subject, and use this to either discover previously used passwords, or gain information about potential passwords our client or target might use for future password-cracking operations. The use of publicly available sources for information gathering is also known as Open Source Intelligence, or OSINT, and we will leverage these sources to try and simplify our work.

In this chapter, we're going to cover the following main topics:

- How does OSINT help with password cracking?
- Leveraging OSINT to access compromised passwords
- Using OSINT to obtain password candidates

How does OSINT help with password cracking?

Open Source Intelligence (**OSINT**) is the craft of obtaining information from publicly accessible sources. While this may sound very straightforward, OSINT constitutes a vast body of tools for finding out information about a person, company, or other subject of interest. We can use this capability to reduce the time and complexity of password cracking in certain situations. How so? You may recall that we talked about the idea of password reuse in the last chapter.

In some scenarios, especially where we require users of a system to create passwords that are too complex to remember (or we require them to rotate these passwords too often), users may *cheat the system* by using the same password for many different systems. As an example, a user might choose a password that is complex but can be remembered, and set up that same password for several systems (for example, websites) and use this solution happily for some time.

Unfortunately, the problem with this approach is that websites, along with other systems, are often compromised for various reasons. When this happens, passwords or password hashes associated with users may be removed from the environment (or exfiltrated) and sold to others or released publicly. Once released publicly, various parties may try to recover the passwords from these information dumps, resulting in large, public tranches of passwords. In some cases, these passwords are associated with a username, such as an email address. In these cases, if the user is reusing passwords across systems, our job of determining their password may be much easier thanks to this public information.

Remember the *RockYou breach* from the previous chapter? This is a great example of a breach whose data has been of great use for password cracking in the years since the breach. But what if a newer breach exposed a password for a user, and they used the same password on other systems? It may now become trivial to recover that user's password without cracking. We continue to revisit the concept of password reuse as it is an important one, because many, many users engage in password reuse. It is even possible that you or a family member has engaged in password reuse at some point, perhaps for less important systems that you do not often use.

As a result, if we know where to look, we can find information on these previously stolen credentials and try them before we engage in more time-consuming cracking operations. Let's start by looking for known, previously compromised passwords.

Leveraging OSINT to access compromised passwords

We need to start this discussion with an important set of caveats about **Operational Security** (**OPSEC**). The idea behind OPSEC is to ensure you avoid the compromise of your personal information, or information related to your systems, when engaging in work online. This section will involve accessing resources on the internet that may be…sketchy in some cases. There are things you can and should do to mitigate the risk of using these kinds of content, and while this section will not cover this in exhaustive detail, it will mention key controls you should consider utilizing. However, you (and only you) are responsible for the security of your own systems!

Many individuals will have many different opinions on this topic – and they all have some validity. Your individual position and risk model may be different. However, once you go beyond some typical internet-facing sites such as `haveibeenpwned.com`, you need to consider what protections you will leverage to secure the environment you are working in from the potential hazards of the areas of the web you visit. The following are just some of the precautions you should consider before moving further along this journey:

- Avoid doing this kind of work on computer systems you use for your normal job. They may contain monitoring enhancements and filtering that can complicate this kind of work, as well as potentially arouse suspicion from your employer. If you are not performing this work on behalf of your employer, you should not use their machines at all.

- Consider using a virtual machine to perform all your browsing and data access work for this objective. This will limit the potential for your host machine to be impacted if you run into malicious content. If you do not have virtualization software, consider free software such as *Oracle VirtualBox*, paired with a free operating system such as one of the various Linux distributions.

- Use a VPN to conceal your internet activity. This can help protect you from being tracked, and also prevent your **Internet Service Provider** (**ISP**) from interfering with your search if you are utilizing websites they consider risky or inappropriate. If you use a VPN, consider a paid solution that does not log your activity, or at least purports not to do so. Also, when using a VPN, consider a different browser from the ones you use for normal internet activity – so for example, if you normally use *Firefox*, consider using *Brave* or another browser.

- Consider using **The Onion Router** (**Tor**) to achieve some of the same objectives as the VPN solutions above. The Tor browser can assist you with this work.

- Use caution when downloading assets (such as files) from these sites. Scan everything you download with at least one, and possibly several, anti-malware products before opening it.

- Use minimal rights on the system you use to access these sites and data and avoid using the `machine/VM` with `admin/root` privileges.

- Be prepared to access this data with command-line tools as many of these files will be very large, and graphical interface tools will likely struggle with their size.

- Some of the searches you will attempt and the resources you may access may contain data that is objectionable to others. Perform this work in a secluded area, not in a cubicle at work.

- Finally, understand the laws of your city/state/country with respect to working with *hacked* data. Much of this chapter focuses on obtaining information on a person's passwords via means other than cracking. However, if you obtain this information, check with local laws and possibly your employer's policies in advance (if it is work-related) to ensure that you are not violating the law nor any organizational policies by searching or downloading breach data. Remember, there are ethical and non-ethical ways to perform this work – and only one of these approaches is supported by this book, and myself.

The first resource we can check for indications of password compromise is one you might have checked personally in the past, *Have I Been Pwned?*, available online at `https://haveibeenpwned.com`. *Have I Been Pwned?* was created by Troy Hunt as a way for users to see whether their email address or phone number – both commonly used as usernames on systems – have been found in known breaches. The data in *Have I Been Pwned?* comes from various sources on the internet, many not accessible to the general public. When appropriate, this data is pulled into *Have I Been Pwned?*, allowing the user of the site to search for breach data related to these identifiers. The interface is fairly straightforward (*Figure 2.1*):

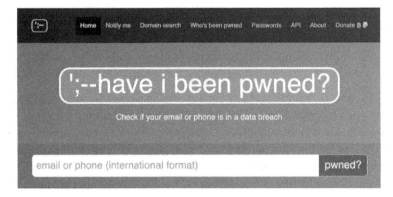

Figure 2.1 – The main page of Have I Been Pwned? as of November 2022

From here we can enter an email address of interest. Bear in mind that email addresses are used as usernames on many different websites, so the compromise of an account associated with this email address may be of interest to us if we can obtain the password.

After entering the email address, press the **pwned?** button to obtain your results as shown in *Figure 2.2*:

Oh no — pwned!
Pwned in 9 data breaches and found 2 pastes (subscribe to search sensitive breaches)

Figure 2.2 – The output of the haveibeenpwned search for the author's email address

You will note that the information returned about the data breaches and pastes is hyperlinked. You can also scroll down on this page to get information about them. Every breach will be different, and the site offers you information about that breach, as well as how to access the data from the breach if possible, as in *Figure 2.3*:

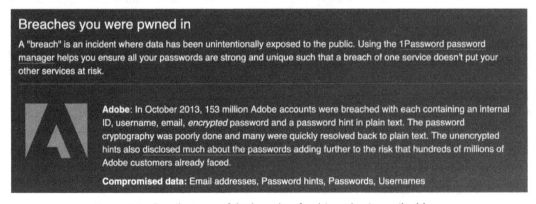

Figure 2.3 – Detail on one of the breaches for this author's email address

In some cases, the links to the breach data may have expired. *Have I Been Pwned?* does not maintain password content itself for ethical reasons. Use caution when accessing breach data links from *Have I Been Pwned?*. Not all internet sites are safe, so consider performing this kind of work in a virtual machine isolated from your host operating system. This data also shows the importance of using different passwords for different sites – a technique we will discuss in more detail towards the end of the book. Also, we can note that many of these breaches contain other personal data, such as password hints that may indicate users' hobbies, family members, etc. These pieces of information can be useful when building a wordlist specific to that particular user or organization as well, especially if we are unsuccessful in obtaining a plaintext password via other means.

Our next stop after using *Have I Been Pwned?* to ensure our subject may have already experienced a breach is *DeHashed*, hosted at `https://www.dehashed.com`. This site maintains records of many compromised assets, and unlike *Have I Been Pwned?*, it also supports searching on many additional fields beyond email and phone number. As we can see in *Figure 2.4*, the interface looks like many websites, with significant information readily accessible:

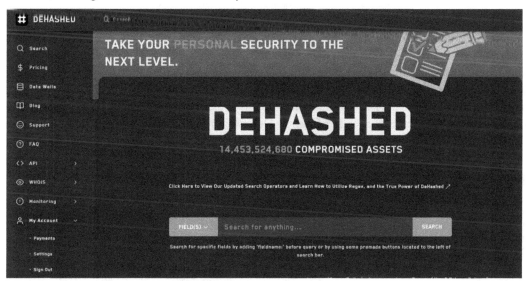

Figure 2.4 – DeHashed after login

Creation of an account at *DeHashed* is free, though accessing detailed search results and the `api` or `whois` records (DNS registrations) does require a subscription plan or the use of credits you can purchase. Using the **Fields** drop-down menu, we can see several options under which to search, as shown in *Figure 2.5*:

Figure 2.5 – List of searchable fields for dehashed.com

In my case, using the same email address entered in *Have I Been Pwned?* yielded a few more results (see *Figure 2.6*), but possibly most interesting was that one result returned an older, actual password from a breach, viewable in cleartext. While payment was required to obtain this data, it shows the power of these kinds of platforms in potentially creating a shortcut and allowing us to obtain passwords via other means. DeHashed is also trying to position itself as a business-friendly platform, so it is possible to request the removal of entries from the DeHashed search results. As a defensive measure, it is worthwhile to review and remove especially sensitive results from these lists, though you may not know if a result is sensitive without paying to view it:

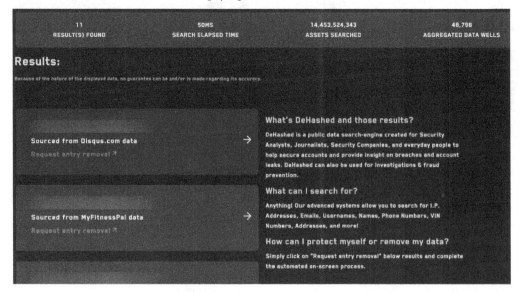

Figure 2.6 – Partial DeHashed search results (email address redacted)

Have I Been Pwned?, *DeHashed*, and other engines may point you to collections of data that contain potential password hashes, or raw data of interest. However, it is often difficult to go to these collections of data, as the data and even the sites hosting it change regularly. In the upcoming section, we will highlight some sites that are currently accessible and have proven helpful in the past. Many files of interest may be accessible via the *deep* (not *dark*) web, namely large or compressed files (or files behind a login) that are not indexed or available in search engines. Again, use caution when accessing these resources further, and perform appropriate digital due diligence. Some resources that have been helpful for this work in the past include the following:

- DDosSecrets (`ddossecrets.com`)
- Searching `Mega.nz leak` on X (formerly Twitter)
- Searching for `combolists`, which are often of varied and undefined structure, but may contain useful information

When working with large data sets such as breach files, some utilities that might be helpful include the following:

- `bulk_extractor` (`https://github.com/simsong/bulk_extractor`)
- Traditional utilities in Linux such as `strings/grep/awk`

Using OSINT to obtain password candidates

It is possible (in fact, somewhat common) that reviewing breach data does not provide us the password we need. So, we may need to resort to cracking to retrieve the data we have been approved to recover. But can we use OSINT to try and make this process more likely to be successful?

Absolutely.

Many people will construct passwords based on things they can easily remember. As we discussed in the previous chapter, mandating password complexity and length can lead users to create easily memorable passwords, often resorting to using words that can be easily recalled. What kind of words are readily recalled? Things we use in everyday life, like the names of our favorite sports teams, our family, our pets, or our hobbies – or even words related to our occupation.

If you are reading this and are blushing because you've created passwords in the past using these kinds of words…don't be embarrassed. We all have done it. But, since our job here is to crack passwords, let's collect the relevant information on our target, and worry about changing our own passwords later! This information can help us create a wordlist that increases the likelihood we can crack the passwords of our target.

To start, we can use the typical search engines and social media sites to obtain information about the interests, family, and work habits of our target.

As you work through the following examples, bear in mind these are not an exhaustive list, and when working through these, note *any* words that are *non-standard* (i.e., industry terms, proper nouns, etc.) and make sure to record these somewhere (possibly in an electronic text file). These will be useful to roll into a wordlist later:

- To research the work history of an individual, we might start with **LinkedIn**, which often contains information about a person's job history, interests, research work, previous companies, and much more. Some basic examples (from real-world situations the author has observed) include the following:

  - Name of the company the target worked at.

  - Key terms for the target's area of study.

  - Cities the target worked in.

  - Terms associated with research work by the target.

  - Terms associated with the target's job (for example, as a scrum master: *standup*, *points*, *backlog*, etc.) These will require a basic understanding of their occupation but are worth it.

- After that, it may be worth reviewing other social media accounts that are less work-related. Many users construct their passwords based on a word, in accordance with their password policies. The question then becomes – what is that word?

  As noted above, it can be work-related. But it can also commonly be related to their family, home, or other things near and dear to the heart of the target.

- For this, we should return to more *traditional* social media sites, such as **Instagram**, **Facebook**, and **X** (formerly Twitter), among others. Search based on the target's account name (often an email address or a username reused across many platforms).

- From here, work through each account, and determine what you can about the target's interests, family members, and more. While this will often require an account on the target application for best success, make note of and catalog your findings. Some examples include the following:

  - Family member names (not just immediate family! Remember grandparents, etc.)

  - Pet names

  - Hobbies of the target

  - Hobbies of family members

- Favorite sports teams of the target

- Favorite sports teams of family members

- Hometown

- Previous schools attended

- Mother's maiden name

- Where the target met their significant other

- Names of previous pets

- Favorite cities or destinations of the target

As a general rule – if a target is willing to talk about a specific person, place, or thing on social media, note it as a possible password.

However, when you obtain potential password candidates, please note these down, ideally in electronic format, as we will try them in future when the time is right. For now, it is time to move on to preparing your wordlists and rules for best cracking effect.

Summary

In this chapter, we discussed using OSINT to potentially identify passwords from previous breaches, to save us the time of identifying and cracking an available password again. However, this approach is not always fruitful, so we discussed obtaining password candidates from other sources as well. Using this information before leveraging cracking can significantly reduce the time and effort required.

Now, after setting up a cracking environment, we need to create appropriate wordlists and rules to use in our various cracking initiatives throughout *Part 2*. This will be covered in more detail in *Chapters 3* and *4* as we discuss the installation of John and hashcat and their associated rules and usage.

3

Setting Up Your Password Cracking Environment

It is possible that after your work in the previous chapters, you have been able to obtain the credentials you need to recover. However, this is not always the case and we need to turn to more overt methods for recovering the credentials we need. You may have also built a supplemental wordlist for use in these processes, which will help us as we move on.

In this book, we will focus on two primary pieces of software for password cracking specifically. It is important to note that we will use other software packages in later chapters to assist with the recovery of certain credential types. In some cases, these will be supplemental programs or scripts to assist in converting hashes to formats recognized by our password-cracking programs. In other cases, we will install software to help us capture the hashes for cracking. We will cover those tools when necessary.

An important thing to note is that all links and URLs are correct at the time of writing this. Things can and do move about the internet at times – luckily, common code repositories such as GitHub serve as a more common place to find needed code in recent years. If links no longer resolve the resource you need, you may need to leverage your search engine of choice to find the content. As we discussed in the previous chapter, choose these resources carefully.

Also, we will assume that you already understand the basics of Linux and Windows filesystems and moving around and running basic commands from the command line.

We will be using **John the Ripper** and **hashcat** as our primary cracking tools throughout this book. Both complement each other in various ways, and each may be the *right* tool for the job in specific circumstances.

In this chapter, we are going to cover the following main topics:

- Installing and introducing John
- Installing and introducing hashcat

Technical requirements

You will need a system (or two systems) on which to run hashcat and John. **Virtual machines** (**VMs**) are useful for documenting and testing software, but they will be limited in the system resources you can use (number of CPU cores and GPUs).

Hashcat will require NVIDIA or AMD drivers for best effectiveness, and these are often easier to install in Windows. As such, you may want to set up hashcat in Windows, which is what we will walk through in this chapter. On the other hand, we will install John on a Linux system, as the installation is more straightforward in that operating system. However, you can absolutely take a different approach. That being said, you will need a system to install these on, and we recommend it be a "real" (not virtualized) system.

Installing and introducing John

John (short for **John the Ripper**) has a long history dating back to the 1990s. The current major version as of this writing is 1.9.0. While this release originally came out in 2019, we continue to frequently see improvements and changes to the code base. However, the team behind John only does releases periodically.

At this point, you have to make a decision – whether you want to use the current release or bring in additional improvements over the past few years via their *bleeding edge* code repository. It is likely that the bleeding edge code repository may have improvements that are not present in 1.9.0; however, the reason things are often referred to as bleeding edge is that you may cut yourself while handling them – in other words, the functionality may be more likely to break or not work in bleeding edge products in general. If you want a more stable experience, work with 1.9.0 if you need the latest and greatest, recognizing that it may include some periodic challenges along the way. In either case, we recommend (and will discuss the installation of) the 1.9.0 *Jumbo* version of this release, which also includes various supporting utilities and capabilities beyond the core release.

You will also need to decide what platform (host operating system) to install John on. In our examples, we will install John on Ubuntu 20.04 LTS. The installation of Ubuntu 20.04 or other Linux distributions is an exercise left to you. To install John 1.9.0-Jumbo, we need to go to the `openwall` GitHub repo and find the release tag for 1.9.0-Jumbo. At the time of this writing, the URL is `https://github.com/openwall/John/releases/tag/1.9.0-Jumbo-1`. This page will give you an option to download the source code for this release, as shown in *Figure 3.1*:

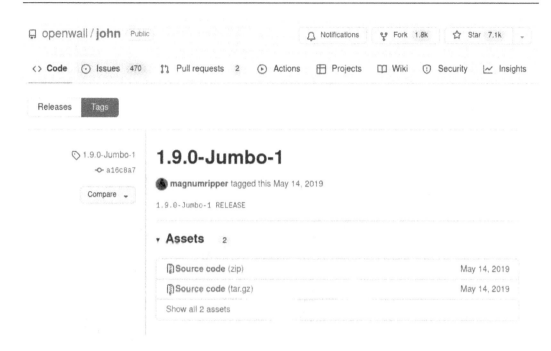

Figure 3.1 – John's 1.9.0-Jumbo-1 release tag in GitHub

Once downloaded, extract the contents of the file; using the built-in archive utility in Ubuntu is fine. We recommend extracting the file to your home directory (/home/username). When this is complete, move into the doc directory under /john-1.9.0-Jumbo-1 and find the install-ubuntu file. This will contain the most current instructions to install John. We will cover the high-level steps here:

1. Ensure that the appropriate build toolchains are installed. At the time of this writing, this is done by running the following command:

    ```
    $ sudo apt-get -y install build-essential libssl-dev git zlib1g-
    dev
    ```

    Note that some installation commands in John will require sudo and some will not; please do not use sudo when not needed as the end product may not work properly.

2. As recommended by the install document, we will install some supplemental packages for performance:

    ```
    $ sudo apt-get -y install yasm libgmp-dev libpcap-dev pkg-config
    libbz2-dev
    ```

3. At this point, install `AMD` or `NVIDIA GPU` support following the instructions.

4. Finally, we are ready to build and install John. From the `/john-1.9.0-Jumbo-1/src` directory, we will run `./configure && make -s clean && make -sj4` (note that this is all one sequence of commands chained together with ampersands (`&&`).

5. Note that this is run without `sudo`. This should ensure that John has everything it needs and we can then build John. In the end, your output should look something like what we see in *Figure 3.2*:

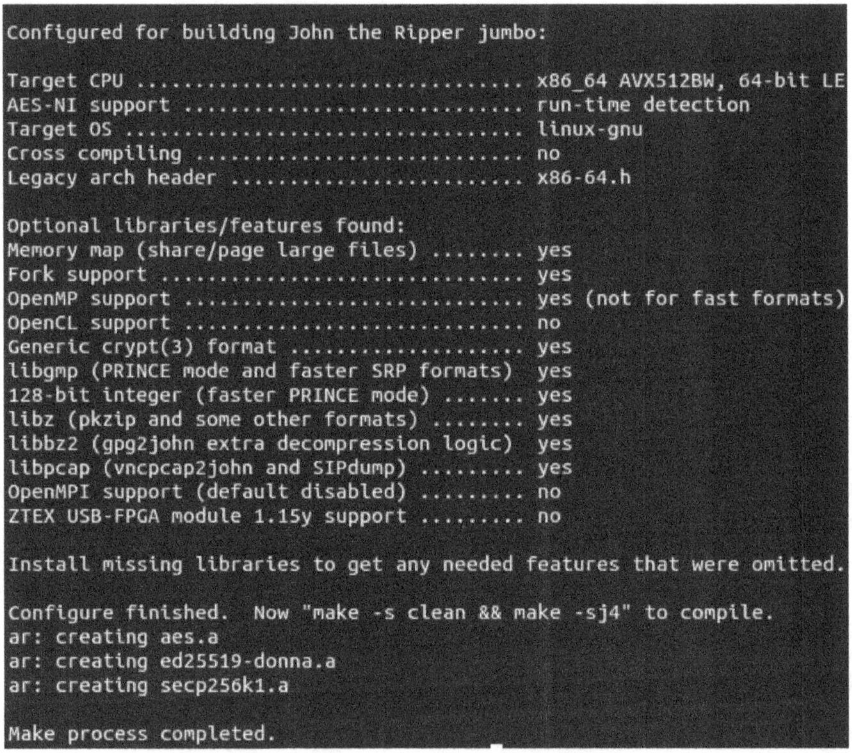

```
Configured for building John the Ripper jumbo:

Target CPU ................................. x86_64 AVX512BW, 64-bit LE
AES-NI support ............................. run-time detection
Target OS .................................. linux-gnu
Cross compiling ............................ no
Legacy arch header ......................... x86-64.h

Optional libraries/features found:
Memory map (share/page large files) ........ yes
Fork support ............................... yes
OpenMP support ............................. yes (not for fast formats)
OpenCL support ............................. no
Generic crypt(3) format .................... yes
libgmp (PRINCE mode and faster SRP formats)  yes
128-bit integer (faster PRINCE mode) ....... yes
libz (pkzip and some other formats) ........ yes
libbz2 (gpg2john extra decompression logic)  yes
libpcap (vncpcap2john and SIPdump) ......... yes
OpenMPI support (default disabled) ......... no
ZTEX USB-FPGA module 1.15y support ......... no

Install missing libraries to get any needed features that were omitted.

Configure finished.  Now "make -s clean && make -sj4" to compile.
ar: creating aes.a
ar: creating ed25519-donna.a
ar: creating secp256k1.a

Make process completed.
```

Figure 3.2 – Successful completion of the John build

Finally, we can test the installation by returning to the /john-1.9.0-Jumbo-1/run directory and running ./john --test=0. You should get an extensive amount of output indicating that it succeeded in all tests and it's ready to go, as shown in *Figure 3.3*:

```
Testing: dynamic_1350 [md5(md5($s.$p):$s) 512/512 AVX512BW 16x3]... PASS
Testing: dynamic_1400 [sha1(utf16($p)) (Microsoft CREDHIST) 512/512 AVX512BW 16x1]... PASS
Testing: dynamic_1401 [md5($u.\nskyper\n.$p) (Skype MD5) 512/512 AVX512BW 16x3]... PASS
Testing: dynamic_1501 [sha1($s.sha1($p)) (Redmine) 512/512 AVX512BW 16x1]... PASS
Testing: dynamic_1502 [sha1(sha1($p).$s) (XenForo SHA-1) 512/512 AVX512BW 16x1]... PASS
Testing: dynamic_1503 [sha256(sha256($p).$s) (XenForo SHA-256) 512/512 AVX512BW 16x]... PASS
Testing: dynamic_1504 [sha1($s.$p.$s) 512/512 AVX512BW 16x1]... PASS
Testing: dynamic_1505 [md5($p.$s.md5($p.$s)) 512/512 AVX512BW 16x3]... PASS
Testing: dynamic_1506 [md5($u.:XDB:.$p) (Oracle 12c "H" hash) 512/512 AVX512BW 16x3]... PASS
Testing: dynamic_1507 [sha1(utf16($const.$p)) (Mcafee master pass) 512/512 AVX512BW 16x1]... PASS
Testing: dynamic_1518 [md5(sha1($p).md5($p).sha1($p)) 512/512 AVX512BW 16x3]... PASS
Testing: dynamic_1528 [sha256($s.$p.$s) (Telegram for Android) 512/512 AVX512BW 16x]... PASS
Testing: dynamic_1529 [sha1($p null_padded_to_len_32) (DeepSound) 512/512 AVX512BW 16x1]... PASS
Testing: dynamic_1550 [md5($u.:mongo:.$p) (MONGODB-CR system hash) 512/512 AVX512BW 16x3]... PASS
Testing: dynamic_1551 [md5($s.$u.(md5($u.:mongo:.$p)) (MONGODB-CR network hash) 512/512 AVX512BW 16x3]... PASS
Testing: dynamic_1552 [md5($s.$u.(md5($u.:mongo:.$p)) (MONGODB-CR network hash) 512/512 AVX512BW 16x3]... PASS
Testing: dynamic_1560 [md5($s.$p.$s2) (SocialEngine) 512/512 AVX512BW 16x3]... PASS
Testing: dynamic_1588 [sha256($s.sha1($p)) (ColdFusion 11) 512/512 AVX512BW 16x]... PASS
Testing: dynamic_1590 [sha1(utf16be(space_pad_10(uc($s)).$p)) (IBM AS/400 SHA1) 512/512 AVX512BW 16x1]... PASS
Testing: dynamic_1592 [sha1($s.sha1($s.sha1($p))) (wbb3) 512/512 AVX512BW 16x3]... PASS
Testing: dynamic_1600 [sha1($s.utf16le($p)) (Oracle PeopleSoft PS_TOKEN) 512/512 AVX512BW 16x1]... PASS
Testing: dynamic_1602 [sha256(#.$salt.-.$pass) (QAS vas_auth) 512/512 AVX512BW 16x]... PASS
Testing: dynamic_1608 [sha256 raw(sha256_raw($p))) (Neo Wallet) 512/512 AVX512BW 16x]... PASS
Testing: dynamic_2000 [md5($p) (PW > 33 bytes) 512/512 AVX512BW 16x3]... PASS
Testing: dynamic_2001 [md5($p.$s) (joomla) (PW > 23 bytes) 512/512 AVX512BW 16x3]... PASS
Testing: dynamic_2002 [md5(md5($p)) (e107) (PW > 55 bytes) 512/512 AVX512BW 16x3]... PASS
Testing: dynamic_2003 [md5(md5(md5($p))) (PW > 55 bytes) 512/512 AVX512BW 16x3]... PASS
Testing: dynamic_2004 [md5($s.$p) (OSC) (PW > 31 bytes) 512/512 AVX512BW 16x3]... PASS
Testing: dynamic_2005 [md5($s.$p.$s) (PW > 31 bytes) 512/512 AVX512BW 16x3]... PASS
Testing: dynamic_2006 [md5(md5($p).$s) (PW > 55 bytes) 512/512 AVX512BW 16x3]... PASS
Testing: dynamic_2008 [md5(md5($s).$p) (PW > 23 bytes) 512/512 AVX512BW 16x3]... PASS
Testing: dynamic_2009 [md5($s.md5($p)) (salt > 23 bytes) 512/512 AVX512BW 16x3]... PASS
Testing: dynamic_2010 [md5($s.md5($s.$p)) (PW > 32 or salt > 23 bytes) 512/512 AVX512BW 16x3]... PASS
Testing: dynamic_2011 [md5($s.md5($p.$s)) (PW > 32 or salt > 23 bytes) 512/512 AVX512BW 16x3]... PASS
Testing: dynamic_2014 [md5($s.md5($p).$s) (PW > 55 or salt > 11 bytes) 512/512 AVX512BW 16x3]... PASS
Testing: dummy [N/A]... PASS
Testing: crypt, generic crypt(3) [?/64]... (2xOMP) PASS
All 407 formats passed self-tests!
```

Figure 3.3 – Output of ./john --test=0

As a last step, we can benchmark John's performance against the various hash types by running `./ john --test` (with no 0). This performs the same self-tests as previously but also does the benchmarking. As a note, this run is quite a bit slower due to benchmarking the 400+ hash types, but this will also provide you with rough data on performance (see *Figure 3.4*):

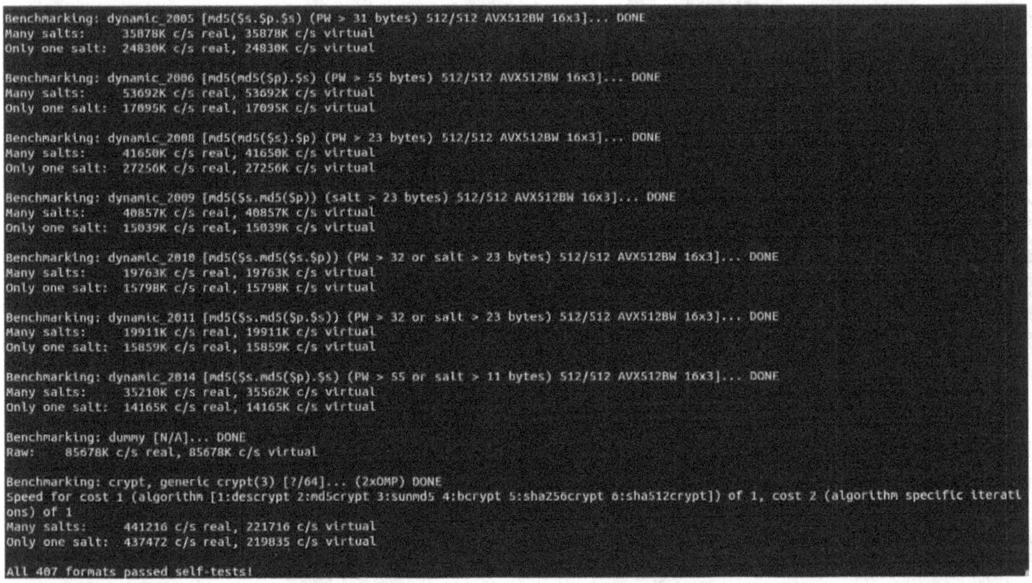

Figure 3.4 – Completed John benchmarking run

So, that's it! We have successfully installed John on our platform and are ready to crack. There was an important side benefit of this installation as well – John's installation also sets up a number of great utility scripts that we can use to help convert recovered hashes to a format that John can use for cracking. This allows John to more readily keep up with file format changes by altering the scripts instead of the core product. If you look at the contents of the `/run` directory where the John binary is located, we can see these varied utilities – either Perl or Python-based – in the same directory (see *Figure 3.5*):

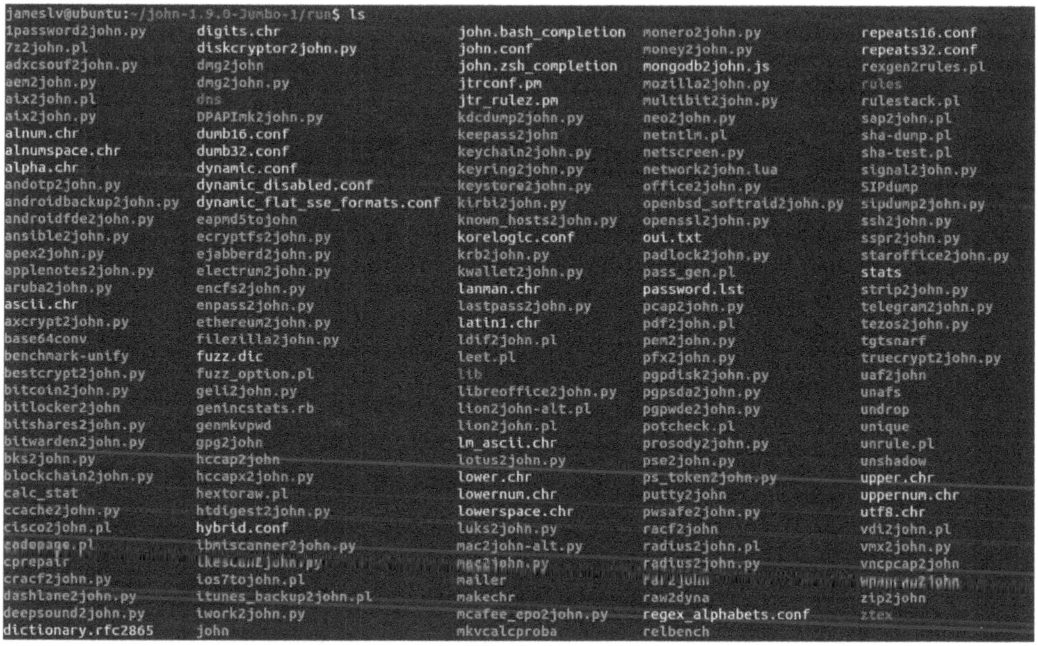

Figure 3.5 – John utilities and other files in the /run directory

These steps have taken you through installing John via the zipped-up source code associated with the 1.9.0 release. If we want the bleeding edge John content, we need to get the source code directories in a different way. We need to prepare a place to put the source, then copy the code down by running the following command in the terminal:

```
git clone https://github.com/openwall/John.git
```

From there, you can follow the previous directions to install the prerequisites, then build and install John. Remember to review the `install` document in the `/doc` subdirectory for any changes.

Now that we have installed John, let us take a quick tour of the core functions of the product.

Core functions of John

John the Ripper offers different modes of operation, each designed for a specific type of password cracking. The primary modes are as follows:

- **Wordlist mode**, which uses a dictionary or wordlist to compare against the hashed password

- **Single crack mode**, which focuses on cracking password hashes with known usernames and other information specific to the user by generating password candidates based on this data – as a result, it is fast but sometimes not successful

- **Incremental mode**, which performs a brute-force attack by systematically generating all possible password combinations within a given keyspace

To select a mode, use the corresponding command-line option, such as `--wordlist` for *Wordlist* mode, `--single` for *Single crack* mode, or `--incremental` for *Incremental* mode.

Word mangling rules in John the Ripper allow you to perform various operations on words from your wordlist to generate new password candidates. These operations can include substitution, deletion, or insertion of characters, capitalization, and more. Word mangling rules help extend the effectiveness of your wordlist and increase the chances of cracking a password. You can create custom rule files or use predefined ones provided with John the Ripper or its community. To use word mangling rules, add the `--rules` option followed by the rule file's name or path.

John also supports character sets to specify the appropriate characters to use while cracking – while English readers may not be familiar with this, other languages may leverage various characters that should be considered in our cracking. Custom character sets in John the Ripper allow you to define the set of characters used to generate password candidates during Incremental mode. By default, John provides several built-in character sets, including lowercase letters, uppercase letters, digits, and special characters. You can also create custom character sets to suit your needs. Custom character sets help limit the search space when you have some knowledge of the password structure, reducing the time required to crack the password.

Installing hashcat

Hashcat has become a great alternative for password cracking work over the past years and excels at GPU-based cracking. While cracking via the GPU cannot be accelerated for all hashing algorithms, hashcat can still introduce significant improvements in many cases.

Where hashcat truly excels is with its ability to leverage GPUs for its various password-cracking activities. The faster the GPU, the faster hashcat can crack passwords. This is the case for both AMD and NVIDIA-based GPUs, though performance on different hashing algorithms can vary significantly. To take advantage of hashcat's strengths, we will need a GPU, which is outside the scope of this book. Given that most individuals will use stronger GPUs for other tasks such as gaming at least periodically, we will perform our hashcat installation on Windows instead of Linux. Thankfully, hashcat provides pre-compiled binaries for the Windows platform.

The binaries for hashcat can be found at `https://hashcat.net/hashcat/`. Here, the binaries for the current version (6.2.6 as of this writing) as well as previous versions can be downloaded. In general, it makes sense to start with the newest release for hashcat – since it is easily installed, it is not terribly difficult to remove it and replace it with an older version if you are experiencing problems. As an additional note, many security professionals express distrust for precompiled software since they cannot sufficiently validate the process by which the software was compiled. While we certainly understand this, in this case, compilation of the hashcat binaries from the source will be an exercise left for you. If you choose to pursue this route (which will also help you understand the product and the process better), refer to the instructions at `https://github.com/hashcat/hashcat/blob/master/BUILD.md`. It is possible to compile natively on Windows with a few different toolchains or compile a Windows build on Linux.

When accessing `https://hashcat.net/hashcat` to download our binaries for Windows, the page will also call out important driver requirements for our cracking hardware (see *Figure 3.6*):

GPU Driver requirements:

* AMD GPUs on Linux require "AMDGPU" (21.50 or later) and "ROCm" (5.0 or later)
* AMD GPUs on Windows require "AMD Adrenalin Edition" (Adrenalin 22.5.1 exactly)
* Intel CPUs require "OpenCL Runtime for Intel Core and Intel Xeon Processors" (16.1.1 or later)
* NVIDIA GPUs require "NVIDIA Driver" (440.64 or later) and "CUDA Toolkit" (9.0 or later)

Figure 3.6 – hashcat driver requirements per https://hashcat.net/hashcat

The path you take here, of course, depends on the hardware of the machine you will be using for cracking. Please install the appropriate software as needed.

Once you download the hashcat binaries, as noted at the top of the page (see *Figure 3.7*), a compressed file will download, likely to your user accounts `/Downloads` directory:

Download

Name	Version	Date	Download	Signature
hashcat binaries	v6.2.6	2022.09.02	Download	PGP
hashcat sources	v6.2.6	2022.09.02	Download	PGP

Figure 3.7 – Current download links from https://hashcat.net/hashcat

Unfortunately, the hashcat binaries package is compressed using 7-Zip, which cannot be opened natively in Windows. If we attempt to open it, we will see that Windows does not know what to do with this file (see *Figure 3.8*):

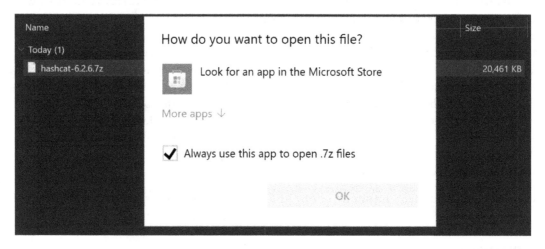

Figure 3.8 – Windows unsuccessfully opening hashcat's 7-Zip package natively

While Windows will prompt us to leverage the Microsoft Store to find the software, in this case, we will leverage the installer provided by 7-Zip, which can be downloaded at https://www.7-zip. org/. This installer is free to use and can be leveraged for multiple file types. When downloading 7-Zip, most modern Windows computers will work best with the 64-bit x64 version (see *Figure 3.9*):

7-Zip is a file archiver with a high compression ratio.

Download 7-Zip 22.01 (2022-07-15) for Windows:

Link	Type	Windows	Size
Download	.exe	64-bit x64	1.5 MB
Download	.exe	32-bit x86	1.2 MB
Download	.exe	64-bit ARM64	1.5 MB

Figure 3.9 – Executable links for 7-Zip from https://www.7-zip.org/

The 7-Zip executable is a self-extracting installer that will ask you to select a destination on the filesystem (the defaults are fine) and then extract and install itself, including some handy additions to the menu options when you right-click on a file. When we right-click on our downloaded hashcat-6.2.6.7z file, we get several options to decompress and extract the file (see *Figure 3.10*):

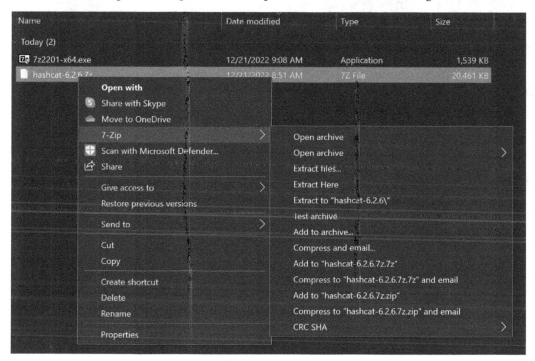

Figure 3.10 – Right-click options for the hashcat 7z file

In this case, let's select **Extract to "hashcat-6.2.6\"**. While this reflects the current release as of this writing, your folder's name may be different based on version. This will create a folder with all the correct hashcat subfolders right in the Downloads directory. We can move this folder anywhere we want on the filesystem. If you would like the "pathed" directory, you can put the folder in an area that supports that or add the folder to the path in Windows. If you do not need this, feel free to move the folder anywhere appropriate for you. Now for the easy part – once you have hashcat extracted, you have everything you need. Open Command Prompt, go to the hashcat folder, then type hashcat --help. If you see a wall of help text, this is a great indicator that hashcat is installed properly and ready to use (see *Figure 3.11*):

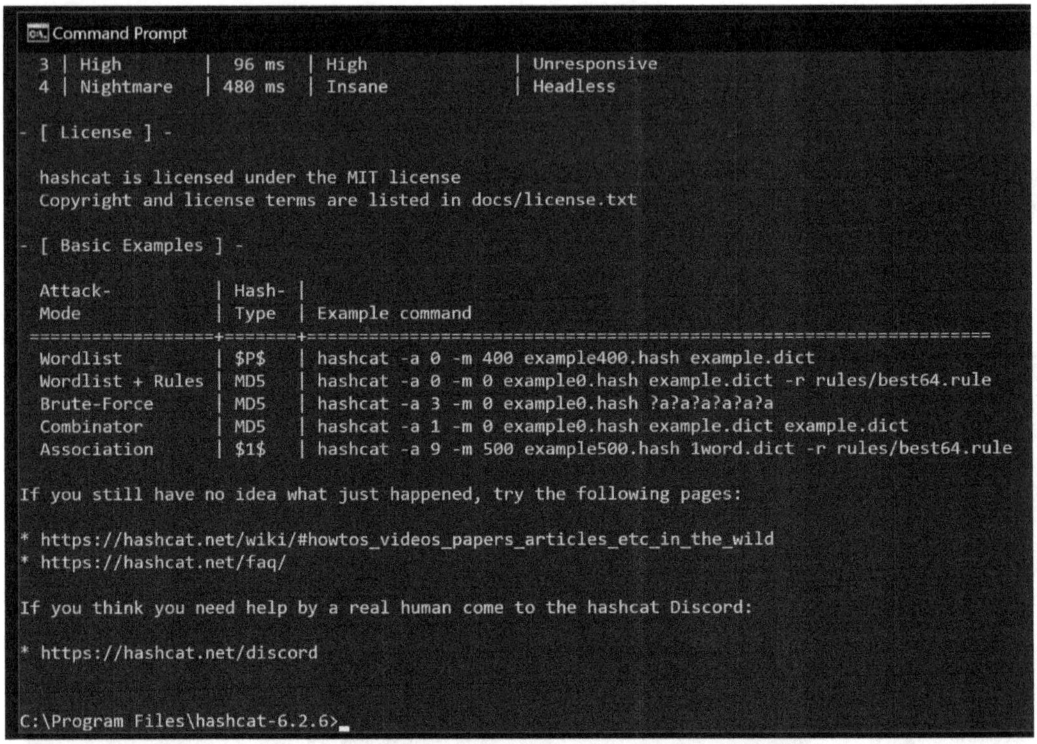

```
[C:\] Command Prompt
 3 | High       | 96 ms   | High      | Unresponsive
 4 | Nightmare   | 480 ms  | Insane    | Headless

- [ License ] -

  hashcat is licensed under the MIT license
  Copyright and license terms are listed in docs/license.txt

- [ Basic Examples ] -

  Attack-       | Hash- |
  Mode          | Type  | Example command
  ================+========+===================================================================
  Wordlist        | $P$   | hashcat -a 0 -m 400 example400.hash example.dict
  Wordlist + Rules | MD5   | hashcat -a 0 -m 0 example0.hash example.dict -r rules/best64.rule
  Brute-Force     | MD5   | hashcat -a 3 -m 0 example0.hash ?a?a?a?a?a?a
  Combinator      | MD5   | hashcat -a 1 -m 0 example0.hash example.dict example.dict
  Association     | $1$   | hashcat -a 9 -m 500 example500.hash 1word.dict -r rules/best64.rule

If you still have no idea what just happened, try the following pages:

* https://hashcat.net/wiki/#howtos_videos_papers_articles_etc_in_the_wild
* https://hashcat.net/faq/

If you think you need help by a real human come to the hashcat Discord:

* https://hashcat.net/discord

C:\Program Files\hashcat-6.2.6>_
```

Figure 3.11 – hashcat help output (several pages long; this is the end of the help content)

Now, let's try a quick benchmark. From the command line, type hashcat -b (see *Figure 3.12*):

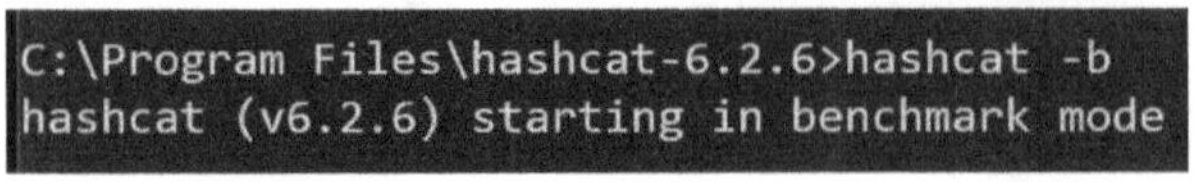

```
C:\Program Files\hashcat-6.2.6>hashcat -b
hashcat (v6.2.6) starting in benchmark mode
```

Figure 3.12 – hashcat starting up benchmark mode

In this mode, hashcat will benchmark *all* of the various hash types supported, one at a time. This is a time-consuming operation but is a good place to find out your overall speed across the various hash types. If you get the first hash type with no errors and your CPU and GPU hardware are represented, this means that your drivers are installed properly. If you did not install your OpenCL and GPU drivers, benchmark mode will likely crash with an error. In our case, we installed OpenCL on this VM and we can see the first hash type tested (MD5) returns a speed and the supported platforms – if more than one platform is supported, such as GPU and CPU, both speeds will be measured and presented (see *Figure 3.13*):

```
OpenCL API (OpenCL 3.0 WINDOWS) - Platform #1 [Intel(R) Corporation]
====================================================================
* Device #1: Intel(R) Core(TM) i7-1068NG7 CPU @ 2.30GHz, 3039/6143 MB (1535 MB allocatable), 2MCU

Benchmark relevant options:
===========================
* --optimized-kernel-enable


--------------------
* Hash-Mode 0 (MD5)
--------------------

Speed.#1.........:   469.8 MH/s (4.31ms) @ Accel:1024 Loops:1024 Thr:1 Vec:16
```

Figure 3.13 – hashcat benchmark output showing hardware platforms and hash rates

If we are interested in learning the hash rate for a specific hash type, we can look up the *mode* associated with that particular hash, then we can benchmark just that one by running hashcat -b -m xxxxx, where xxxxx is the particular mode to test. In this example, we will benchmark mode 22000 for cracking WPA2-PSK passphrases for wireless using hashcat -b -m 22000 (see *Figure 3.14*):

```
C:\Program Files\hashcat-6.2.6>hashcat -b -m 22000
hashcat (v6.2.6) starting in benchmark mode

Benchmarking uses hand-optimized kernel code by default.
You can use it in your cracking session by setting the -O option.
Note: Using optimized kernel code limits the maximum supported password length.
To disable the optimized kernel code in benchmark mode, use the -w option.

OpenCL API (OpenCL 3.0 WINDOWS) - Platform #1 [Intel(R) Corporation]
====================================================================
* Device #1: Intel(R) Core(TM) i7-1068NG7 CPU @ 2.30GHz, 3039/6143 MB (1535 MB allocatable), 2MCU

Benchmark relevant options:
===========================
* --optimized-kernel-enable


----------------------------------------------------------------
* Hash-Mode 22000 (WPA-PBKDF2-PMKID+EAPOL) [Iterations: 4095]
----------------------------------------------------------------

Speed.#1.........:    17363 H/s (29.08ms) @ Accel:1024 Loops:1024 Thr:1 Vec:16

Started: Wed Dec 21 09:54:41 2022
Stopped: Wed Dec 21 09:54:48 2022
```

Figure 3.14 – hashcat benchmark results for mode 22000

Let's take a quick look through some of the hashcat features and capabilities.

Core functions of hashcat

Hashcat offers different attack modes, each designed for a specific type of password cracking. There are three main modes: Straight mode, which uses a dictionary or wordlist to compare against the hashed password; Combination mode, which combines two dictionaries to create new password possibilities; and Brute-force or `mask attack` mode, which generates all possible password combinations within a given keyspace. To select a mode, use the `-a` flag followed by the mode number (0 for Straight, 1 for Combination, or 3 for Brute-force). Additionally, options exist to combine a mask and wordlist (and vice versa) as well as association mode, which is similar to the single crack mode in John (using some previously known data on the user such as the username).

Hashcat also supports rules. Rules in hashcat allow you to perform various operations on words from your dictionary to generate new password candidates. These operations can include the substitution, deletion, or insertion of characters, capitalization, and more. Rules are a powerful way to extend the effectiveness of your dictionary and increase the chances of cracking a password. You can create custom rule files or use predefined ones provided by the hashcat community. To use rules, add the `-r` flag followed by the rule file's path. We will talk about rules in more detail in *Chapter 4*.

As mentioned earlier, hashcat supports mask attacks, a type of brute-force attack. Masks in hashcat are patterns that define the structure of the password candidates generated during Brute-force mode. Masks consist of placeholders, which are replaced with characters from a predefined or custom character set. They help you limit the search space when you have some knowledge of the password structure, reducing the time required to crack the password.

With this, our John and hashcat installations are ready to go!

Summary

In this chapter, we installed John and hashcat. While these are not the only software components we will use over the course of this book, these are the foundation for the rest of our success, and we will deal with other installations as needed in other chapters.

In the next chapter, we will review rule construction to increase the probability of successful password cracking.

In *Part 2*, we will start addressing the cracking of common hash types, chapter by chapter, so feel free to jump around to the chapters that are directly relevant to your current cracking work to find out more.

4
John and Hashcat Rules

As we have seen, John and hashcat can be used to perform various types of cracking attacks against credentials. However, brute-force style or mask attacks can be overly time-consuming, and wordlist-based (or dictionary) attacks may result in fewer cracked credentials when they are not present in the wordlist.

To try and move in between these two extremes, we can use **rules**, which are a way of taking a source list of candidates for cracking (such as a wordlist) and modifying those candidates to increase the likelihood of successful cracking. These modifications can be simple, such as capitalizing the first character of a candidate or adding a number to the end of a candidate. On the other hand, we can also engage in significant substitutions from the original candidate.

As an important note, all links and URLs are correct at the time of writing. Things can and do move about the internet at times – luckily, common code repositories such as GitHub have meant that there's a more common place to find the necessary code in recent years. If links no longer resolve the resource you need, you may need to leverage your search engine of choice to find the content. As we discussed in *Chapter 2*, choose these resources carefully.

Also, we will assume that you already understand the basics of Linux and Windows filesystems and moving around and running basic commands from the command line.

In this chapter, we are going to cover the following topics:

- Analyzing password complexity rules
- Selecting and using John rules
- Selecting and using hashcat rules

Analyzing password complexity rules

Before we start using rules to enhance our password cracking, we need to take a step back and analyze the target system we are going to crack credentials for. The reason for this is straightforward – if we know what a credential *requires*, we can start to immediately *include* and *exclude* certain types of credential constructions and formats.

As an example, many organizations will follow the logic of complexity over length. As we discussed in *Chapter 1*, this is not necessarily the best approach, but since it is still often used, it requires examination. A common method of setting credential requirements to reflect complexity over length would be to require three (or four) of the following four criteria:

- Lowercase letters
- Uppercase letters
- Numbers
- Special characters (such as hyphens (-), exclamation points (!), and so on

In addition to the preceding requirements, we may also require users to rotate (or change) credentials every 90 days or so.

Humans – the species that, for the moment, is most commonly responsible for password selection – will generally choose credentials that are easy for them to remember. As mentioned in *Chapter 1*, this will result in a few common approaches to constructing credentials in light of the preceding requirements. Let's start by looking at the requirement of three of four character types. Many individuals will select *uppercase*, *lowercase*, and *numbers* for these three options, simply because they are easier to remember than a special character. Unfortunately, with these choices, many humans will quickly fall into a predictable pattern when selecting their credentials.

In the English language, the first letter of a proper noun, as well as the first letter of a sentence, is often capitalized. Because these occurrences are so frequent, English speakers will often select a credential and capitalize the first character of the credential because this is easy for them to remember and meets one of the complexity requirements. Following the first character is the rest of the credential, often represented in lowercase characters, which is in keeping with normal English writing style. Now, with this, the user has met two of the complexity requirements – but they need at least one more! Where can they add a number or special character? Unfortunately, the easy path for this user at this point will be to append the number to the end of the original credential. So, for example, let's say that a user is creating a new credential and decides to base it off a dictionary word – in this case, *computer* (another common technique of credential creation will result in the user looking around their desk for something to use as the base credential, often resulting in office items being a key part of their password). Applying the aforementioned logic, we can take the base word of *computer* and meet two of the three complexity requirements by capitalizing the first letter – that is, *Computer*. Great! But we need one more complexity requirement to be met. Maybe we could add a number to the end of the credential – that would be easy to remember. How about *Computer1*?

Now, let's take a moment to reflect. Is this a *good* credential, meaning one that is resistant to cracking? Not really. It is based on a dictionary word (computer) and uses common substitutions for some of the characters. But is this a credential that meets the requirements of the organization? Yes, it is. With this, we can begin to see how complex credential requirements do not necessarily protect users' credentials more effectively.

What if the user elects to choose a special character instead of a number? The user will likely append the special character to the end of the base word instead of the number. If both special characters and numbers are required, the user will often append one, then the other, to the end of the base word. These are all scenarios we can readily account for with rules, as we will see later.

While this already seems problematic, this method of constructing credentials to meet common requirements lends itself to a sinister method of credential construction, leading to very predictable passwords. This is especially true when the preceding complexity requirements are coupled with a 90-day password rotation, meaning the user must select a new password every 90 days. With these requirements, and a 90-day rotation, a user can construct a password they can *easily* remember. What changes every 90 days? While you might not always think so when you look at the weather report, the seasons change every 90 days, and now some very simple, very memorable passwords leap to mind that you can easily rotate at the next password change.

For example, is it spring? Well, if so, we can make the base word spring and capitalize it to get *Spring*. But we need to add a number or a special character. Wait… what year is it? *Spring23* or *Spring2023* are passwords that meet complexity requirements and are easy for the user to remember. When the user changes passwords 90 days later, they may elect for *Summer23* or *Summer2023*, and likewise for fall (or Autumn) and winter.

Now, to be clear, *none of these are "good" passwords*. They are predictable and based on dictionary words, and we can readily build rules to crack credentials such as these. However, if these requirements were not in place, we wouldn't necessarily want to make cracking along these sequences of characters our priority.

How can we determine what password construction rules a target site, domain, or something else uses? That is as easy as creating an account. Most authentication systems will tell you their complexity requirements when you create an account. If you cannot create an account, try using an existing username and triggering the **Forgot Password** workflow, which may provide information on the password's complexity requirements. Also, note the *maximum* credential length on your target – many authentication systems will limit passwords to 20 characters, or even as short as 14 or 16 characters, making some brute-force or mask-style attacks more feasible, depending on how the credential is stored.

Now that we have done some research on the specific requirements for credential construction, let's look at how we can construct or use existing rules in hashcat and John to simplify this process.

Selecting and using John rules

Like hashcat, John has a tremendous amount of support for creating and using custom rules to interact with our candidates to increase the chances of a successful crack. The number of options will easily become overwhelming. In many cases, it may be better to start with a sample file with some common rules, alongside code comments that make it clear what a particular rule does.

> **Note**
>
> There are great resources on the internet for this topic. Refer to https://www.openwall.com/john/doc/RULES.shtml, as well as https://www.openwall.com/john/doc/EXAMPLES.shtml, for some additional guidance.

These documents can also be found locally under the /doc directory, where John was cloned, as we described in *Chapter 3*. See *Figure 4.1* for a list of the current documentation in john 1.9.0-Jumbo-1:

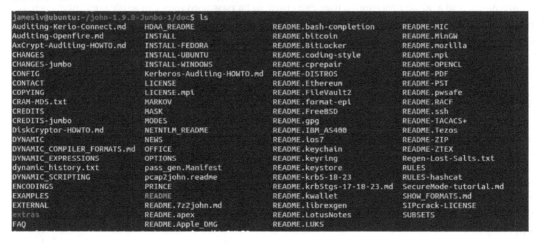

Figure 4.1 – The list of documentation installed alongside John

Rules are set up and configured for repeated use in the john.conf file. This file is typically located in the /run directory, below the destination directory where you cloned John. Opening this file with a text editor yields a lot of settings, as shown in *Figure 4.2*:

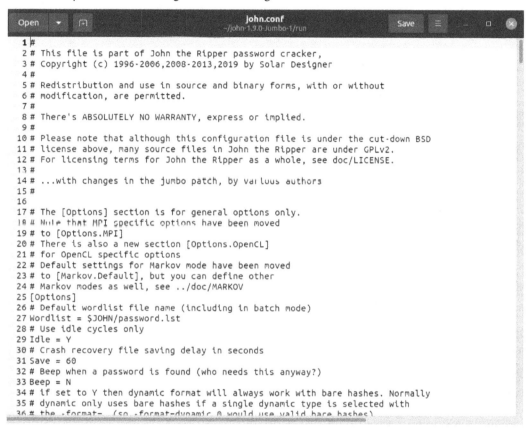

Figure 4.2 – The john.conf file in a text editor

This file contains a lot of settings, not just those regarding password rules. We can see the password rules by running find on the rules string, as shown in *Figure 4.3*:

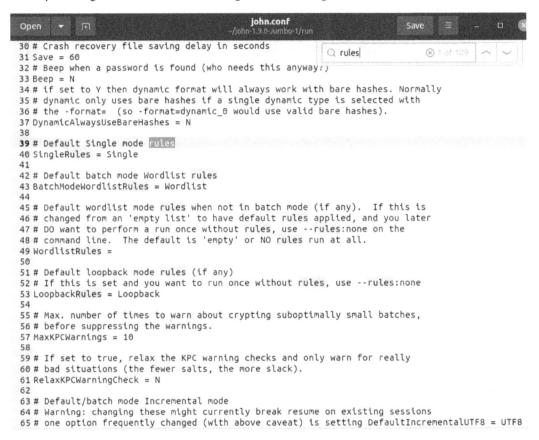

Figure 4.3 – Searching for the "rules" string in john.conf

There are a lot of hits here! We can further restrict this by enhancing our search to find the List. Rules phrase, as shown in *Figure 4.4*:

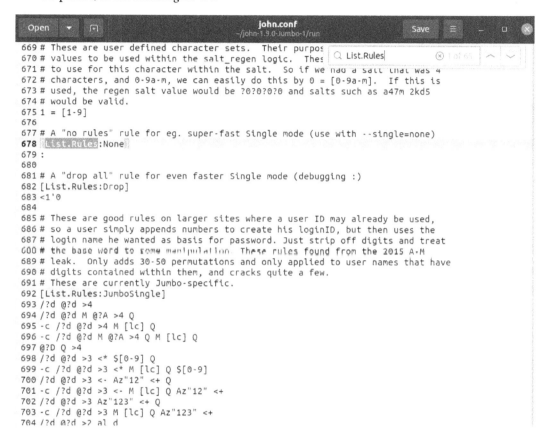

Figure 4.4 – Searching for the rules in john.conf

We're getting somewhere interesting! We can look at the rules that have already been configured for John to understand how they work and as a precursor to building our own.

Let's search for a rule that has a lot of variability and is well-commented so that we can see what's going on. Change your search so that it looks for the List.Rules:Wordlist string, as shown in *Figure 4.5*:

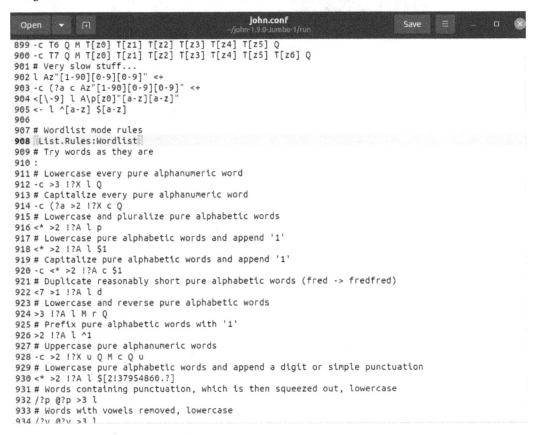

```
Open    ▼    ⊞              john.conf                Save    ☰    _    ☐    ⊗
                        ~/john-1.9.0-Jumbo-1/run
899 -c T6 Q M T[z0] T[z1] T[z2] T[z3] T[z4] T[z5] Q
900 -c T7 Q M T[z0] T[z1] T[z2] T[z3] T[z4] T[z5] T[z6] Q
901 # Very slow stuff...
902 l Az"[1-90][0-9][0-9]" <+
903 -c (?a c Az"[1-90][0-9][0-9]" <+
904 <[\-9] l A\p[z0]"[a-z][a-z]"
905 <- l ^[a-z] $[a-z]
906
907 # Wordlist mode rules
908 List.Rules:Wordlist
909 # Try words as they are
910 :
911 # Lowercase every pure alphanumeric word
912 -c >3 !?X l Q
913 # Capitalize every pure alphanumeric word
914 -c (?a >2 !?X c Q
915 # Lowercase and pluralize pure alphabetic words
916 <* >2 !?A l p
917 # Lowercase pure alphabetic words and append '1'
918 <* >2 !?A l $1
919 # Capitalize pure alphabetic words and append '1'
920 -c <* >2 !?A c $1
921 # Duplicate reasonably short pure alphabetic words (fred -> fredfred)
922 <7 >1 !?A l d
923 # Lowercase and reverse pure alphabetic words
924 >3 !?A l M r Q
925 # Prefix pure alphabetic words with '1'
926 >2 !?A l ^1
927 # Uppercase pure alphanumeric words
928 -c >2 !?X u Q M c Q u
929 # Lowercase pure alphabetic words and append a digit or simple punctuation
930 <* >2 !?A l $[2!37954860.?]
931 # Words containing punctuation, which is then squeezed out, lowercase
932 /?p @?p >3 l
933 # Words with vowels removed, lowercase
934 /?v @?v >3 l
```

Figure 4.5 – Looking at the wordlist rules that are there by default in john.conf

As shown in the comments, this rule is intended for use in wordlist mode, meaning that wordlists will be used as potential candidates for cracking. But these rules will *mangle or modify* the candidate from the wordlist and perform various permutations on it to increase the probability of a successful crack. So, how do these manglings or modifications work? Let's dissect one rule as an example; you can work through the rest yourself if you need to. We will highlight the rule on line 922; the comment on line 921 explains what the rule does.

The rule is `<7 >1 !?A l d`. Let's use the documentation at `https://www.openwall.com/john/doc/RULES.shtml` to determine what's going on here, and break it down piece by piece:

- `<7`: This is a length control command and ensures this rule is only used if the candidate is less than seven characters long.

- `>1`: This is another length control command and ensures this rule is only used if the candidate is more than one character long. So, this rule will only be used for a candidate that's two to six characters in length.

- `!?A`: This will reject a candidate (`!?`) that does not match a specific character set (in this case, `A` is the opposite of `a`, which is an alphabetic character). Essentially, this rule will not be performed on a candidate that contains anything other than alphabetical characters.

- `l`: This will convert the candidate into lowercase.

- `d`: This will duplicate the candidate. In other words, `fred` will become `fredfred`.

A lot is going on here. Again, read the comments, and certainly look at more of them to see how these are constructed. If we want to leverage these rules (which are meant to be used with a wordlist), we can call these rules with `–rules=rule_name`, where `rule_name` is the identifier after the word `Rules` in the `john.conf` file. In this case, we would call this block of rules with `–rules=Wordlist`.

We can continue to review the `john.conf` file for other, pre-created rules and use these as needed, or build new ones. Now, let's look at hashcat's rules engine.

Selecting and using hashcat rules

Hashcat rules use a similar construction format to John rules and are called from the command line, just like John rules are. However, hashcat rules are often standalone as a separate file, and hashcat rules are more readily able to be contributed to the community, as opposed to us having to integrate the changes into the `john.conf` file in John.

With its focus on GPU-accelerated cracking, hashcat expects a constant stream of input to ensure the GPU is fully utilized for maximum efficiency. As such, hashcat rules tend to be a bit more aggressive than some of the John rules we have seen.

Hashcat rule construction is well defined on the following web page at the time of writing: `https://hashcat.net/wiki/doku.php?id=rule_based_attack`.

Common readily available hashcat rules are included with the hashcat repository and can be cloned from the hashcat GitHub repository at `https://github.com/hashcat/hashcat`.

Choosing which hashcat rules you should use is about striking a balance between time and ability. Hashing algorithms where a candidate can be calculated quickly tend to lend themselves to more rules to allow for greater coverage and potentially more successful cracking. On the other hand, slower algorithms might be better with fewer rules to balance coverage and time more effectively.

While several common rules are included with hashcat if it's installed via GitHub, some of the more common ones are noted here. At the time of writing, all of these rules have variations in the GitHub repository that have been contributed by the community:

- Best66: This is a commonly used rule, and its name stems from the fact that it originally had 64 rules, though the set has grown beyond that at this point. Best66 was designed to be a balanced ruleset and covers common password construction such as word reversal, appending of one or two digits to a password, some leetspeak substitutions (replacing e with 3 and similar substitutions), as well as some basic appending and prepending of characters.

- d3ed0ne: This rule, named after the handle of the creator, leverages a significantly greater ruleset (over 30,000 lines at the time of writing). The d3ed0ne rules focus on a lot of appending (up to four numeric characters in some cases) as well as character position swapping, duplication of characters, and more.

- T0X1C: Again, this is named after its creator. T0X1C has over 4,000 rules that cover a broad range of password construction scenarios, including truncation, appending, reversal, and more. This is a balanced but longer list than Best66.

- Dive: Also named after its creator, dive contains a lot of rules (almost 100,000) but focuses on fairly quick transformations, repeats of characters, full repeats of the candidate, and more. It is surprisingly fast for its length.

- Leetspeak: This is named after the tendency to substitute common alphabetical characters with numbers. One of the more recognizable leetspeak substitutions may be the use of 1337 or 31137 to substitute for elite. This rule is small and will produce approximately 17 substitution-based passwords for each candidate.

While it may be daunting to look at the password rules in hashcat and understand what they are doing, it may be easier to run a sample wordlist through hashcat while leveraging a rule and looking at its output.

To do this, create a text file containing your wordlist. For speed and simplicity, try just one word, such as password. In both Windows and Linux, this can be done with the echo password > word command. While we chose word as the filename, you can choose whatever you like, though our screenshots will continue to reference word.

To perform a test run of a rule, try the following command:

```
hashcat -r <rulename> --stdout word > <rulename>.output
```

In this example, replace <rulename> with the name of the rule you wish to use, such as dive. rule. The -stdout switch sends the output of the rule to the screen, and we redirect it with > to our output filename so that we can easily review it in an editor.

Let's see what the `dive.rule` output looks like. Let's run the rule, as shown in *Figure 4.6*:

```
hashcat -r D:\Desktop\hashcat-6.2.6\rules\dive.rule --stdout word.txt > dive.output
```

Figure 4.6 – The hashcat command in Windows to run the dive rule against a test wordlist

Figure 4.7 shows the output of the dive rule against our simple candidate of `password`:

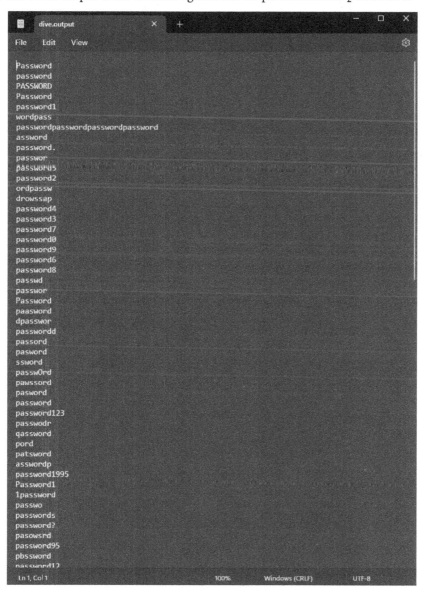

Figure 4.7 – Opening the output file of our test run in an editor

`dive` took our single password candidate of `password` and created over 99,000 possible passwords to try in just a couple of seconds. However, think about how this scales against a wordlist of thousands or tens of thousands of candidates, which is why the speed of GPU-aided cracking is so important. Are all of these permutations valuable in our cracking operations? That will depend on the target's password policies, as well as other factors, and why deep rules may not be valuable in all cases.

We can take hashcat even further and combine two rules. This needs to be done with caution as it may overwhelm the memory available to hashcat. To show an example of this, we will run the `dive` and `leetspeak` rules together against our test password of `password`. When rules are run together, every line of each rule is compared against every line of the other rule. This means our output of potential candidates should be `X*Y`, where `X` is the number of lines in the first rule and `Y` is the number of lines in the second rule, assuming only one candidate is calculated per line.

The command to run this is shown in *Figure 4.8*:

```
hashcat -r D:\Desktop\hashcat-6.2.6\rules\dive.rule -r D:\Desktop\hashcat-6.2.6\rules\leetspeak.rule --stdout word.txt > leet+dive.output
```

Figure 4.8 – The command to run to use both the "leetspeak" and "dive" rules against our test candidate

Figure 4.9 shows the output of the `leetspeak` and `dive` rules in our text editor:

Figure 4.9 – Output of the "leetspeak" and "dive" rules together in a text editor

As we can see, the two rules have produced over 1.6 million candidates from our source candidate of `password`. While it does not perfectly match up to X*Y, it is pretty close, meaning we likely have a few lines in `dive` that have multiple rules in them. Now, consider that using these two rules together against a 10,000 candidate wordlist would result in hashcat performing 16,845,000,000 candidate hashing and comparisons. This will mean significantly more time for the cracking operations – how much more time will vary depending on the specific algorithm and how many guesses per second your specific hardware platform can do.

Summary

In this chapter, we introduced you to rules for both John and hashcat, which allow us to take a list of candidates and perform substitutions and manipulations that are commonly performed when users are forced to choose passwords that are, by their requirements, hard to remember. Carefully selecting the right rules for a cracking operation can make a huge difference in the amount of credentials recovered.

In *Part 2*, we will learn about cracking specific types of hashes, how they work, how they are constructed, how to retrieve them, and how to crack them. The chapters in *Part 2* are designed to be more or less standalone – you do not need to read through all of them at once, just reference the ones you need as you need them. With that, let's start discussing specific hash types.

Part 2:
Collection and Cracking

In this part, we will dive into specific hash types and how to collect them, how to format them, and how to crack them. While this will not be an exhaustive treatment of hash types, we will cover some of the most prominent ones you may need to interact with, and the steps can be substituted for other hash types with some modifications.

This part has the following chapters:

- *Chapter 5, Windows and macOS Password Cracking*
- *Chapter 6, Linux Password Cracking*
- *Chapter 7, WPA/WPA2 Wireless Password Cracking*
- *Chapter 8, WordPress, Drupal, and Webmin Password Cracking*
- *Chapter 9, Password Vault Cracking*
- *Chapter 10, Cryptocurrency Wallet Passphrase Cracking*

5

Windows and macOS Password Cracking

The two most prominent desktop operating systems at this point are, by far, some variation of Windows or macOS. Windows has held a dominant position in the desktop space for years and cemented that hold with versions such as Windows 95, 98, XP, and, more recently, Windows 7, 10, and 11. These systems often store credentials locally for logging in users, even if the network is not available. If we can retrieve the hashes of these passwords, we can attempt to crack them to recover access or whatever other needs are required.

On the other hand, once an operating system frequented by artists and executives, macOS has found an increasingly common home in the corporate office space, and it's also used by many home users. While originally Unix-based, macOS (the name for the Apple desktop operating system, not to be confused with iOS or iPadOS, which run on other hardware platforms) has seen significant change since its earliest days, including how it stores and accesses secrets on the device.

In this chapter, we will cover the following topics:

- Collecting Windows password hashes
- Cracking Windows hashes
- Collecting macOS password hashes
- Formatting/converting macOS hashes into their expected formats
- Cracking macOS hashes

Before we begin, here is a note about Windows passwords.

> **Note**
>
> Depending on your desired outcome, it may be possible to interact with and authenticate to Windows systems by simply providing the hash of the user, even without knowledge of the password. This is referred to as a pass-the-hash attack or relaying attack, and they remain effective to this day.

Assuming this is not what you are looking for, we will proceed to hashes and cracking.

Collecting Windows password hashes

Windows password hashes take a few different forms. The two hashes that are stored on almost every Windows device to authenticate a local user are the **LAN Manager** (also known as **LANMAN**) hash and the NT hash.

The LANMAN hash represents – unfortunately – some of the worst password hashing that can be available in a modern operating system. The good news is that LANMAN hashing is disabled by default in newer Windows operating systems (Windows 7 and higher). However, it is worth discussing LANMAN because it is still enabled on some systems for backward compatibility, and it makes cracking passwords monumentally easier. Why?

It has to do with how the original password is stored and treated. As you will remember from *Chapter 1*, the more characters available to a given position in the password, the greater the possibilities for the correct character for that position – also, the longer the password, the greater the number of combinations possible for the password. LANMAN hashes actively undermine both of these principles. LANMAN hashing starts with a password of 14 characters or less (LANMAN was designed for older systems dating back to the 1980s when shorter passwords were quite common). From there, LANMAN would take the password and perform the following actions:

- Convert the password into all uppercase characters (this reduces the potential characters from 52 to 26 for each position).

- To work with a uniform-length password, LANMAN would then pad the password with NULL characters until it was 14 characters in length.

- LANMAN would then split the password into two seven-character parts. This is probably the single greatest error in the LANMAN algorithm. This means that instead of cracking a 14-character password, we are cracking two seven-character passwords. Assuming a character set of alphanumeric (36 characters – remember, everything is uppercase), a 14-character password would be 36^{14} or 6,140,942,214,464,815,497,216 possibilities. By splitting the password into two parts, it is now $36^7 + 36^7$ or 156,728,328,192 possibilities. That is an enormous simplification of the work to be done in cracking.

- Two DES keys are created from these seven-character parts and are used to encrypt a known string of data.

From a security perspective, the good news here is that LANMAN hashes are disabled by default in modern Windows systems. If LANMAN hashing is disabled on a system, the LANMAN hash will be replaced with `aad3b435b51404eeaad3b435b51404ee`, which indicates a blank LANMAN hash. The overall `hashdump` format for a Windows system will look like this:

```
johndoe:1000:aad3b435b51404eeaad3b435b51404ee:NT hash:::
```

The first value, `johndoe`, is the user account name. `1000` is the **relative identifier** (**RID**) of the account on that system, which is followed by the LANMAN hash (blank in this case) and the NT hash. It is important to recognize LANMAN hashes and their overall weaknesses because some companies will enable LANMAN hashes even on newer Windows systems for various business reasons. If so, cracking becomes markedly easier.

The NT hash represents a significant improvement over the LANMAN hashing process and is stored locally for use with local logins on the Windows device. The NT hash takes a different approach to hash calculation as opposed to LANMAN:

- The user's password is converted into Unicode (allowing for preservation of case, along with non-English character sets).
- The Unicode-converted password is then hashed using the MD4 algorithm. This will produce a fixed-length output of 128 bits, or 16 bytes, in length.
- This MD4 hash output is then stored for future authentication.

Unfortunately, a problem with both of these hashes is the lack of a **salt** or an additional bit of random data that's included with each hash calculation. The value of a salt is that it can thwart **rainbow table** attacks, where an attacker can precompute the hashes for a large number of possible passwords. Because salts are unknown, this prevents precomputation.

In addition to the LANMAN and NT hashes, which are stored locally, Windows also uses certain types of hashes to authenticate a user over the network. These hashes are typically referred to as NTLMv1, NTLMv2, and Kerberos. Let's talk about each in detail.

NTLMv1 hashes authenticate the user using a **challenge/response** process to validate the user, which uses the user's credentials to authenticate the user. However, no salt (again) is used when going through the following process:

1. **Client credential request**:

   The client requests access to a server resource.

   The server generates a 16-byte random number known as the *challenge* and sends it to the client.

   I.   Client response:

   - The client retrieves the user's NT hash from its local database (this hash is generated from the user's password)

   - It uses this NT hash to encrypt the server's challenge

   - The client sends this encrypted challenge (known as the *response*) back to the server

2. **Server verification**:

   I.   The server forwards the client's response, along with the original challenge and the username, to the domain controller.

   II.   The domain controller uses the username to retrieve the corresponding NT hash and encrypts the challenge.

   III. If the domain controller's encrypted challenge matches the client's response, the user is authenticated.

   IV.   For cracking, this is known as `netntlmv1`, and it can be found as a separate mode in hashcat.

NTLMv2 hashes represent a significant evolution of this protocol as they go through the following process:

1. **Client credential request**:

   Similar to NTLMv1, the client requests access, and the server sends back a 16-byte challenge.

   I.   Client response:

   - The client generates a *client challenge*, which is a random number (this adds variability and resistance against certain attacks)

   - The client also gets the current time and includes this in the response (timestamp), further adding to the response's uniqueness

   - The client then creates an HMAC-MD5 hash of these values (server challenge, client challenge, timestamp, and some additional information) using the NT hash of the user's password

2. **Server verification**:

    I.   The server sends the response to the domain controller with the username and the original challenge.

    II.   The domain controller, using the stored NT hash and the same data (challenges, timestamp), calculates the HMAC-MD5 hash.

    III. If the hashes match, the client is authenticated.

This protocol is referred to as `netntlmv2` in hashcat for cracking purposes. In both scenarios, over-the-network authentication captures interesting data that's potentially useful for cracking.

Finally, the Kerberos protocol may yield some interesting hashes as it uses the existing protocol to access the existing network resources.

Kerberos is a network authentication protocol that's designed to provide strong authentication for client/server applications using secret-key cryptography.

Kerberos

Developed at MIT, Kerberos is named after the mythological guardian of the underworld. While used in many networks, we are interested in its use in Windows Active Directory domains.

The following are key Kerberos components:

- **Key Distribution Center (KDC)**: This is a central authority that provides temporary session keys and tickets. It consists of two parts: the **Authentication Server (AS)** and the **Ticket Granting Server (TGS)**.

- **Client and server**: The device requesting and granting access, respectively.

- **Authentication process**:

  - Initial authentication:

    - The client requests an authentication token (also known as a TGT) from the AS in the KDC.

    - The client proves its identity to the AS, typically using a password.

    - The AS responds with the TGT and a session key, both encrypted with the client's password.

  - TGT and service request:

    - The client decrypts the TGT using its password, and then requests access to a specific service from the TGS, using the TGT.

    - The TGS validates the TGT and issues a service ticket, encrypted with the service's secret key.

- Service authentication:

  - The client sends the service ticket to the desired service.

  - The service decrypts the ticket using its secret key and verifies the client's credentials.

  - If the credentials are valid, the service grants access to the client.

- **Usage**:

  - Kerberos is widely used in secure environments, especially in enterprise settings (not just Windows!).

  - It's integral to Windows Active Directory, providing the necessary authentication mechanism for logging into Windows domains.

- **Advantages and considerations**:

  - It provides strong, scalable authentication

  - It relies on synchronized time among all entities (clients and servers)

  - It requires a centralized KDC, which can be a single point of failure

Kerberos is considered more secure than NTLM due to its reliance on mutual authentication and encrypted communication. It's particularly effective in environments where strong, centralized authentication management is necessary. However, it does require network components in the KDC and time servers to work properly. There are opportunities to grab tickets and crack them to retrieve the user's password as well!

Cracking Windows hashes

Cracking operations on Windows hashes depend significantly on the hashes that are observed, and this is most easily supported by hashcat. The modes vary, depending on the type of hash involved. They are listed here:

- **LANMAN hashes**: Mode 3000

- **NTLM hashes**: Mode 1000

- **Netntlmv1**: Mode 27000

- **Netntlmv2**: Mode 27100

- **Kerberos**: Varies

The best way to ensure your hashcat hash is formatted properly is to check the hashcat example hashes page at `https://hashcat.net/wiki/doku.php?id=example_hashes`.

After formatting your hash, pass it to hashcat with the proper attack mode and hash type, along with your wordlist or mask as appropriate. The following example syntax is for `ntlm`, hash mode `3000`:

```
hashcat -m 3000 -a 0 ntlm.hash rockyou.txt
```

Remember that, due to their relatively simple construction, LANMAN and NTLM hashes may not need to be cracked and may be able to be looked up, especially since these hashes are not salted. One resource for looking up NTLM hashes is `https://ntlm.pw`, which has over 8 billion unique hashes and can look through these very quickly. A sample output of this site is shown in *Figure 5.1*:

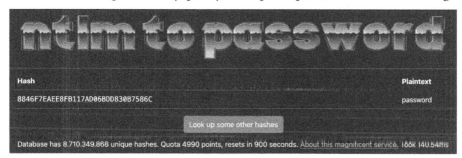

Figure 5.1 – ntlm.pw looking up our ntlm hash of "password"

In this example, we obtained an `ntlm` hash, but you can also generate an NTLM hash for a given password using utilities such as the wonderful CyberChef at `https://gchq.github.io`. Upon feeding that hash into `https://ntlm.pw`, it was able to find the hash associated with the password of `password` in less than a second. With over 8 billion hash entries on this site and a relatively small lookup time, it is worth trying this before performing a more time-consuming cracking operation.

Now, let's pivot to macOS passwords and discuss those in greater detail.

Collecting macOS password hashes

macOS stores secrets in a different manner than other operating systems we have seen to date, and as such, an explanation of how these secrets are stored is needed.

In macOS, anything that would be considered a *secret* in the operating system is stored in a file called the **keychain**. This file is essentially a store for all the secrets that are protected by macOS. This can include, but is not limited to, the following:

- Network passwords
- Application passwords
- Form fill for Safari (web browser)

- Cryptographic keys
- Wi-Fi passwords
- Trusted certificate authorities

While all this is very interesting, the focus of this section is on the macOS password, also known as the login password, for macOS – the password the user enters to sign into the system itself. The reason the macOS keychain is relevant to this objective is because *the user login password is used to encrypt the keychain*. This means that if we can successfully guess the user's login password, we can successfully decrypt the keychain, which tells us what the user's password is. As a bonus, we can also have access to the keychain contents itself, which can contain many other secrets that may be valuable to us.

> **Note**
>
> The preceding description is the default behavior for macOS. It is possible to set a different password for the keychain that differs from the login password. However, this requires an overt choice to do this and is extremely rare in practice; as such, the following techniques will work fine in almost all situations, but understand that some small edge cases exist.

The macOS keychain for the user is located in the filesystem at `/Users/username/Library/Keychains`. While there may be more than one keychain, `login.keychain-db` is likely to be the right target. In the case of our test system, we can see multiple `login.keychain-db` files, but the one with the current date is the correct target, as noted in *Figure 5.2*:

```
Keychains $ cd /Users/jameslv/Library/Keychains/
Keychains $ ls -l
total 848
drwx------  14          staff     448 Aug  8  2021 3F8BA19C-AD18-547C-A0EE-AA95D276C3D0
drwx------  11          staff     352 Jun 10 09:45 B91AAB89-37F4-5688-B6A3-EDA914B6EE6A
-rw-r--r--@  1          staff   20460 Jun 16 07:52 Microsoft_Entity_Certificates-db
-rw-r--r--@  1          staff  387840 Jun 21 01:22 login.keychain-db
-rw-r--r--@  1          staff       0 Aug 11  2021 login.keychain-db.sb-ef5cf4ed-r5gmgC
-rw-------   1          staff   23804 Jun 10 09:47 metadata.keychain-db
```

Figure 5.2 – Keychains on a macOS Monterey (12.6) system

Let's try using `file` command to see if there's anything special about the file structure of the keychain by issuing the `file login.keychain-db` command, as shown in *Figure 5.3*:

```
Keychains $ file login.keychain-db
login.keychain-db: Mac OS X Keychain File
```

Figure 5.3 –"file" command run against the macOS keychain

Well, wasn't that interesting? The keychain appears to be a macOS proprietary format. The `file` command reads the Magic Bytes at the beginning of the file, which identifies what type of file it is. We will use this capability to our advantage later when running the `keychain2john` utility.

Looking more closely at *Figure 5.2*, you may also notice that the keychain can be read by all users. This means we can actually remove the keychain to another location, or even another machine.

We will likely want to move the file to another machine for cracking, so let's take a copy of this file and get it over to our cracking setup, as shown in *Figure 5.4*::

```
Keychains $ cp login.keychain-db /Users/      /crackme.keychain
Keychains $ ls -l /Users/        /crackme.keychain
-rw-r--r--@ 1          staff  387840 Jun 21 01:26 /Users/        /crackme.keychain
```

Figure 5.4 – Copying the keychain file over to a new file using the macOS "cp" command

As we can see, we were able to successfully copy the keychain file to an area of the filesystem we had permissions on, specifically within the `/Users` area of the filesystem associated with the logged-in user. As noted previously, it is important to copy the keychain file using the `cp` command, not move the keychain file using the `mv` command, so that the original remains in place.

Now that we have the keychain file in our possession, it should be moved to our cracking platform by whatever means are appropriate for our solution.

Formatting/converting hashes into their expected formats

The macOS keychain is encrypted and not readily able to be worked with in its existing format. This means that we need to convert the keychain file so that we can extract the hash and crack it appropriately. John can do this in a standalone utility called `keychain2john.py`. This utility is installed when you install John with the jumbo patch, as we defined previously.

Once `keychain2john.py` is installed, we can readily extract the contents needed from the keychain. Let's go through the `keychain2john` script and see some of the basics of what's happening here.

At the time of writing, `keychain2john` can be found at `https://github.com/openwall/john/blob/bleeding-jumbo/run/keychain2john.py`. `keychain2john` performs the following core functions:

- It scans the keychain for the appropriate magic number
- It extracts key items such as the salt, IV, and encryption key

`keychain2john` requires `python3` and takes one argument: the keychain file to extract key data from.

> **Note**
>
> If you get any errors when running `keychain2john.py`, install the `python-is-python3` package on Linux, which will point `python2` calls to `python3`. `python2` is not supported on modern Linux systems.

`keychain2john` will look through the file for the magic header of `fade0711` in hex format. This header indicates the start of the `db` blob within the keychain, which should be located at the end of the file. Once that magic header has been found, `keychain2john` will extract specific elements and print them out to the screen in `Filename:$keychain$*salt*iv*ciphertext` format.

> **Note**
>
> The filename will be recognized by John, but this particular element will need to be removed if we're cracking this with hashcat. The `$keychain$` string will identify the hash type to the cracker; salt, IV, and ciphertext are self-explanatory. All of these will be separated by asterisks.

In our example, taking the user login keychain from our macOS device, and moving it to our cracking machine, then running the keychain through `keychain2john.py`, produces the results shown in *Figure 5.5*:

```
    /@ubuntu:~$ /home/_      /john/run/keychain2john.py crackme.keychain
crackme.keychain:$keychain$*15ddadf53a2ad6495b9444f2237          *a132016c5b0143ba*      bf4d55b
cb88e255b7e81697754733cd4077021cdd45ab0f0ac649f42337ede484ffb4f946162133b2a
```

Figure 5.5 – Running keychain2john.py against our test macOS
keychain (with salt and ciphertext redactions)

From here, we can push this output into a file using the > operator in Linux, as shown in *Figure 5.6*, to produce a file for cracking with John:

```
@ubuntu:~$ /home/      /john/run/keychain2john.py crackme.keychain > mykeychain.hash
@ubuntu:~$ ls -l mykeychain.hash
-rw-rw-r-- 1           183 Jun 20 22:46 mykeychain.hash
```

Figure 5.6 – The output of keychain2john.py being pushed into the mykeychain.hash file

With this, we now have a macOS hash to use for cracking.

Cracking hashes

Once we have extracted the hash from the keychain, we can use John or hashcat to crack this hash. John will use the `$keychain$` string in the `keychain2john` output to help it understand the correct hash type. From there, we just need to supply a wordlist, rules (if desired), and the hash, as shown in *Figure 5.7*:

```
      @ubuntu:~$ /home/        /john/run/john --wordlist=rockyou.txt mykeychain.hash
Note: This format may emit false positives, so it will keep trying even after finding a possible candidate.
Using default input encoding: UTF-8
Loaded 1 password hash (keychain, Mac OS X Keychain [PBKDF2-SHA1 3DES 512/512 AVX512BW 16x])
Will run 4 OpenMP threads
Press 'q' or Ctrl-C to abort, 'h' for help, almost any other key for status
              (crackme.keychain)
1g 0:00:02:26 DONE (2023-06-20 22:44) 0g/s 97963p/s 97963c/s 97963C/s !!123sabi!!123..*7¡Vamos!
Session completed.
```

Figure 5.7 – John being run against the keychain hash from our target system

In this example, we have used a modified version of the RockYou wordlist for demonstration purposes. Also, note that the way that keychain hashes are checked during the cracking operation can result in false positives (essentially, the compare operation is looking at a 4-byte padding value, which is small enough that hash collisions can arise); as such, we want to ensure that we completely exhaust our wordlists, which John will do by default.

To perform the same cracking in hashcat, we need to use mode 23100. Unlike John, we will need to direct hashcat to exhaust the wordlist by passing --keep-guessing to hashcat at the command line.

If you try to use the same hash file in hashcat that you used in John, hashcat will fail because it is not expecting the filename piece in the keychain2john output. Open up the hash file, and remove the filename so that the $keychain$ string is at the beginning of the file, as shown in *Figure 5.8*:

```
        @slingshot:~/Desktop$ cat mykeychain.hash
$keychain$*15ddadf53a2ad6495b9444f22375d7e3fa8ac658*a13201     *ff1beaa69b6010bf4d55bcb88e255b7e
81697754733cd4077021cdd45ab0f0ac649f42
```

Figure 5.8 – hash file modified by removing the filename at the
beginning to ensure that it can be used by hashcat

From here, we can run hashcat with a wordlist-based attack (-a 0) against a keychain hash (-m 23100), and ask hashcat to exhaust the wordlist (--keep-guessing), as shown in *Figure 5.9*:

```
        slingshot:~/Desktop$ hashcat -m 23100 -a 0 mykeychain.hash rockyou.txt --keep-guessing
hashcat (v6.2.5) starting

OpenCL API (OpenCL 2.0 pocl 1.8  Linux, None+Asserts, RELOC, LLVM 11.1.0, SLEEF, DISTRO, POCL_DEBUG) - Platform #1 [The pocl project]
=============================================================================================================================
* Device #1: pthread-Intel(R) Core(TM) i7-1068NG7 CPU @ 2.30GHz, 1439/2942 MB (512 MB allocatable), 2MCU

Minimum password length supported by kernel: 0
Maximum password length supported by kernel: 256

Hashes: 1 digests; 1 unique digests, 1 unique salts
Bitmaps: 16 bits, 65536 entries, 0x0000ffff mask, 262144 bytes, 5/13 rotates
Rules: 1

Optimizers applied:
* Zero-Byte
* Single-Hash
* Single-Salt
* Slow-Hash-SIMD-LOOP

ATTENTION! --keep-guessing mode is enabled.
This tells hashcat to continue attacking all target hashes until exhaustion.
hashcat will NOT check for or remove targets present in the potfile, and
will add ALL plains/collisions found, even duplicates, to the potfile.
```

Figure 5.9 – hashcat instantiation and startup

In this example, there were no hash collisions, and as such hashcat showed an **exhausted** status when the wordlist was complete, but the hash was cracked much earlier. While this was printed out on the screen, it may not be apparent. We can verify the cracked passwords by running hashcat again with the - - show option, specifying our target hash file, as shown in *Figure 5.10*:

```
         slingshot:~/Desktop$ hashcat -m 23100 mykeychain.hash --show
$keychain$*15ddadf53a2ad6495b9444f22375d7e3fa8ac658*a13201        *ff:    cd4077021cdd45ab0f0ac649f42337e
de484ffb4f946162133b2a:
```

Figure 5.10 – The output of the hashcat –show command for the target hash. The cracked password is displayed at the end of the hash, after the colon (redacted here)

At this point, we can use this hash to access the information stored within the keychain, as well as potentially log in as the user, if they have not set a different password for logging in and the keychain.

Summary

In this chapter, we discussed how to identify, extract, and crack Windows and macOS password hashes. With these, you will be able to successfully recover passwords on the majority of common desktop operating systems. In the next chapter, we will discuss a similar tactic for a growing server operating system – Linux.

6

Linux Password Cracking

For many years, the Unix and Linux **operating systems (OSs)** have been the backbone of the infrastructure of many companies, as well as the internet as a whole. Unix and Linux are sometimes referred to interchangeably due to the similarities between many of the OS components. However, at their core, Unix is an older OS that has typically been a proprietary, licensed solution, while Linux evolved as a free and open source OS. This does not mean there are no versions of Linux that you pay for; rather, these are often for support and maintenance purposes. As such, while this chapter will typically refer to Linux, most of this guidance can be used with Unix as well.

In this chapter, we're going to cover the following main topics:

- Collecting Linux password hashes
- Formatting/converting hashes into their expected formats
- Cracking hashes

A note about Linux passwords

Depending on your objective and the reason for password cracking, you may just want to gain access to the target device, and not need to know the user's password. In these cases, we can use a feature of Linux called *single-user mode* to boot the OS and reset the password of the root (administrator) user to a known value. Once the administrator password is known, we can reset any user password we choose on the system.

To boot into single-user mode, follow these instructions for Ubuntu 20.04 while using *systemd* for system management. However, note that these directions may vary depending on the type and version of the system. Research how to boot into single-user mode for your system before proceeding:

1. On boot, use the *Esc* key to access the bootloader.

2. Select the top option in the boot menu and hit the *E* key to edit the pre-boot elements.

3. One of the last lines should start with the word "Linux" and a path to the kernel (typically in the /boot partition). Go to the end of this line and add systemd.unit=rescue. target.

4. Press *F10* to boot the system in single-user mode.

5. Once booted, the system output should say something like "Press enter for maintenance." Pressing *Enter* at this point will bring you to a Terminal prompt with root privileges.

6. From here, you can reset the password of any user to a value of your choosing. Use the passwd command to reset the password of the root user, or directly reset a user password by typing passwd <username>, replacing <username> with the name of the target user.

7. Once your recovery is complete, you should be able to type exit at the prompt to reboot the system into normal (multi-user) mode.

As you can see, with physical access to the system, recovering access is straightforward. However, this cannot recover passwords for you, so whether you can use this shortcut depends on your objectives.

Collecting Linux password hashes

To obtain the password hashes for Linux, we will need to be able to run commands on the target Linux device. The means to obtain the ability to run commands on the device are outside the scope of this book and can be more broadly covered in other books that focus on penetration testing techniques. These techniques can include, but are not limited to, exploiting a service or running process on the system, accessing a user's account and logging into the system via normal channels, and running commands remotely via injection flaws, web shells, and similar modes of access. Again, these are outside the scope of this book and are thoroughly covered in other resources. For now, we will assume you have access to a user account on your machine and show output as it would be presented in the Terminal.

Linux password hashes are commonly represented in a file typically located at /etc/shadow. Access to this file is restricted to the root user or someone who can run commands as the root user via utilities such as sudo. The output of an attempt to access the /etc/shadow file with normal user privileges is shown in *Figure 6.1*:

```
@ubuntu:~$ cat /etc/shadow
cat: /etc/shadow: Permission denied
```

Figure 6.1 – The output of attempting to access /etc/shadow with insufficient privileges

However, if we run the same command again with sudo, assuming we have access to do so, we will get the output of the /etc/shadow file, as shown in *Figure 6.2*:

```
@ubuntu:~$ sudo cat /etc/shadow
root:!:19068:0:99999:7:::
daemon:*:18858:0:99999:7:::
bin:*:18858:0:99999:7:::
sys:*:18858:0:99999:7:::
sync:*:18858:0:99999:7:::
games:*:18858:0:99999:7:::
man:*:18858:0:99999:7:::
lp:*:18858:0:99999:7:::
mail:*:18858:0:99999:7:::
news:*:18858:0:99999:7:::
uucp:*:18858:0:99999:7:::
proxy:*:18858:0:99999:7:::
www-data:*:18858:0:99999:7:::
backup:*:18858:0:99999:7:::
list:*:18858:0:99999:7:::
irc:*:18858:0:99999:7:::
gnats:*:18858:0:99999:7:::
nobody:*:18858:0:99999:7:::
systemd-network:*:18858:0:99999:7:::
systemd-resolve:*:18858:0:99999:7:::
systemd-timesync:*:18858:0:99999:7:::
messagebus:*:18858:0:99999:7:::
syslog:*:18858:0:99999:7:::
_apt:*:18858:0:99999:7:::
tss:*:18858:0:99999:7:::
uuidd:*:18858:0:99999:7:::
tcpdump:*:18858:0:99999:7:::
avahi-autoipd:*:18858:0:99999:7:::
usbmux:*:18858:0:99999:7:::
rtkit:*:18858:0:99999:7:::
dnsmasq:*:18858:0:99999:7:::
cups-pk-helper:*:18858:0:99999:7:::
speech-dispatcher:!:18858:0:99999:7:::
avahi:*:18858:0:99999:7:::
kernoops:*:18858:0:99999:7:::
saned:*:18858:0:99999:7:::
nm-openvpn:*:18858:0:99999:7:::
hplip:*:18858:0:99999:7:::
whoopsie:*:18858:0:99999:7:::
colord:*:18858:0:99999:7:::
geoclue:*:18858:0:99999:7:::
pulse:*:18858:0:99999:7:::
gnome-initial-setup:*:18858:0:99999:7:::
gdm:*:18858:0:99999:7:::
sssd:*:18858:0:99999:7:::
jameslv:$1$LkF5gpwu                       :19068:0:99999:7:::
systemd-coredump:!!:19068::::::
```

Figure 6.2 – The /etc/shadow file when read with sudo

Let's unpack the contents of this file. Each line represents a different user account. There are eight pieces of information in each line, and they are separated by the colon (:) character. These eight pieces of information are as follows:

- **Username**: The first field at the left of the line, up to the first colon, is the username. This is the name the user will utilize to log in to the machine.

- **Password hash**: The second field is the password hash. You will notice that on all the accounts except the `jameslv` account, the password hash is represented as a `*` or a `!` character. This means that these users have no password set. This also means these users cannot typically log in with a password-based login. You will notice, however, that the `jameslv` user has a (partially redacted) password hash. This hash contains multiple pieces of information (more on that in a moment).

- **Last password change**: The third field indicates the day the password was last changed. This doesn't note the date itself, but rather the number of days that have passed since `January 1, 1970` – also known as the *Unix epoch*. This date was selected by Unix engineers in the 1970s to represent when time – as a Unix system knows it – began. Linux maintains the same format.

- **Minimum password age**: The fourth field represents how many days must pass after setting a password before the user can change it. Note that in this case, the minimum password age is `0`, which means a user on this system can rotate a password at any time.

- **Maximum password age**: This is how long a password can be used before it must be changed. The value of `99999` here means that the passwords on this system essentially never expire.

- **Password expiration warning**: The sixth field indicates how many days before password expiration the user will begin to receive warnings that they need to change their password.

- **Inactivity period**: The seventh field indicates how long the account can remain on the system without being used before it is disabled.

- **Account expiration date**: Linux accounts can be set to expire on a pre-defined date. For example, if you have a contractor whose engagement ends on a specified date, you can set that in advance here.

Let's dig into that password hash field (*field 2*) in more detail since a lot is going on here. This field follows the following format: `$id$salt$hash`.

Three pieces of information are in this one field. The first piece, the *ID*, specifies the type of hashing algorithm being used. Linux (and Unix) support multiple types of algorithms for password hashing, and you can have different passwords for different users with different hashing algorithms being used in the same `/etc/shadow` file. The common hashing algorithms in Linux are as follows:

- `$1$`: MD5

- `$2$`: Blowfish (there are several variations of Blowfish, so this may be represented by `$2a`, `$2b`, and so on)

- `$5$`: SHA-256

- `$6$`: SHA-512

The second portion of the field is the **salt**. The salt is a bit of random data that's added to the calculation of the password hash. This means that a salted hash will have a different hash output than an unsalted hash, which has dramatic security benefits. A traditional hashing algorithm will produce the same hash text whenever it's presented with the same input. For example, if I hash the phrase `password` with MD5, the hash output will be `5f4dcc3b5aa765d61d8327deb882cf99`. If I hash the phrase `password` with MD5 on your computer, or mine, or any other computer out there on the internet, this will be the hash output. In the world of password cracking, this consistency causes two problems:

- If I review a `/etc/shadow` file and see multiple password hashes that are identical, this means their passwords are identical as well. Cracking or obtaining one password will yield the same password for all the accounts with the identical hash.

- An attacker could choose to pre-calculate the MD5 hashes for a given wordlist of password candidates. This would greatly reduce the time to perform password cracking, at the expense of the storage and time to pre-calculate and store those password hashes. These types of storage arrays are known as **rainbow tables** and can shortcut the cracking process.

To mitigate these issues, we can *spice up* our password hashes by adding this salt – a randomly selected data value – and storing it alongside the password hash. This salt is not a secret – if you have access to the shadow file, you can view it – but it mitigates the attack scenarios mentioned previously.

The third portion of this field is the password hash itself. This is the result of the salt and the password being run through the algorithm specified in the first part of the field.

When a user attempts to access the system, this hash is recalculated and compared against the password hash here, using the salt and the algorithm specified. If the hashes match, the authentication is successful; if not, it is rejected.

Not all hashing algorithms in Linux and Unix are created equally – which is why they support many different hash types. They allow us to migrate users over time to newer and more secure hashing algorithms.

Looking at *Figure 6.2*, you will note that the `jameslv` account was hashed using MD5, which is considered an inferior algorithm at this point, though salting does help. Let's look at this from a defensive perspective for a moment and see how we can set this configuration on a Linux machine.

In a base Linux system, password hashing algorithms will be set in a **Pluggable Authentication Module (PAM)** configuration file. On this `Ubuntu 20.04` test virtual machine, this is located in `/etc/pam.d/common-password`. Accessing this file requires elevated privileges, so we will need to access this from the Terminal with `sudo nano /etc/pam.d/common-password`, as shown in *Figure 6.3*:

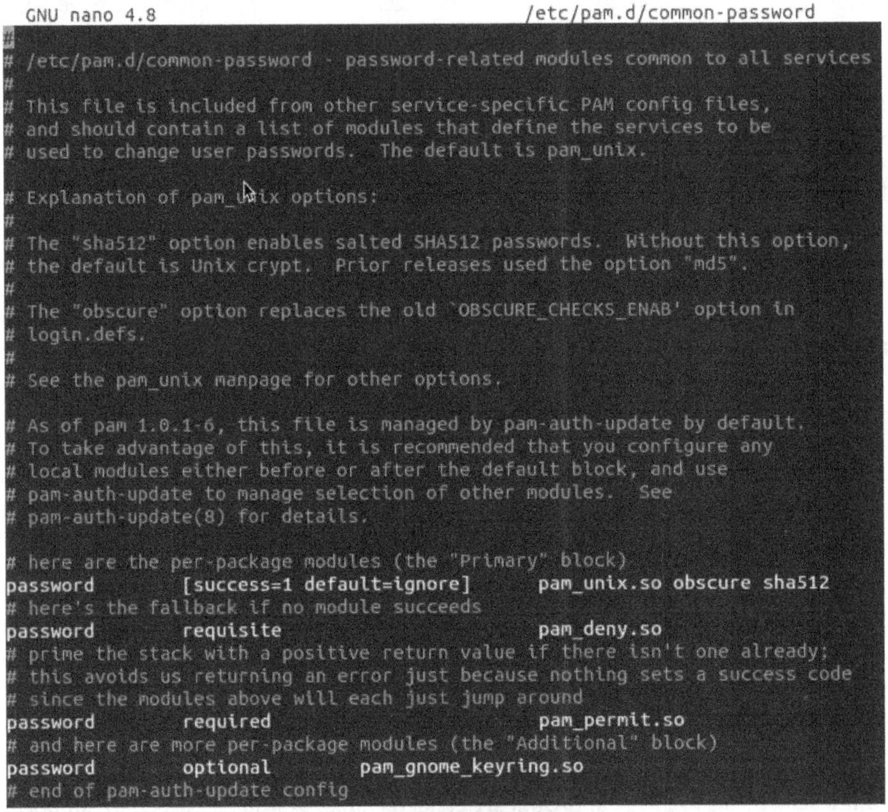

Figure 6.3 – Password configuration on Ubuntu 20.04

As we can see, the password configuration should require a `sha512` hashed password for users, as noted in the line with `pam_unix.so obscure sha512`. Let's test this by creating a new user on this system, which should leverage a `sha512` hash according to these settings. We can do this by issuing `sudo adduser test` at the Terminal to create a new test user, as shown in *Figure 6.4*:

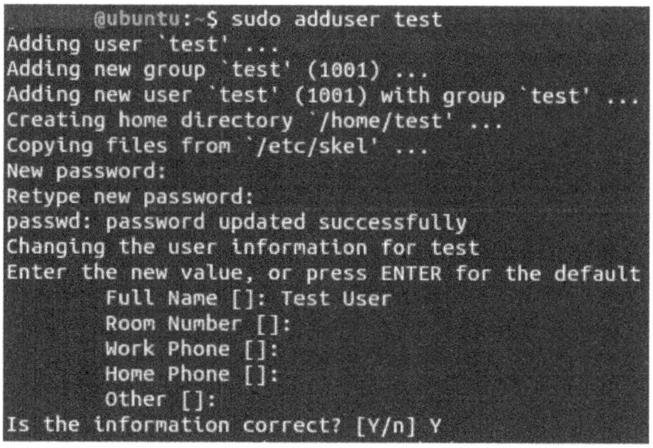

Figure 6.4 – Creating a new user on Ubuntu 20.04

You may have noticed the questions about the full name, room number, and other details, and that we haven't seen that information yet. We will come to that shortly. Now, let's look at the /etc/shadow file again and see what data is in there for our new user:

Figure 6.5 – The /etc/shadow file after the new test user has been created

Looking at this file, we can see a few differences. First, we can see that the new test user has had a password hash created using SHA-512, as evidenced by 6 at the beginning of the second field in the /etc/shadow file, as opposed to the MD5 hash that was created for the james1v user. Additionally, both accounts were created with the same password, which has created very different hashes between the two algorithms (and different salts selected by the system; MD5 uses an 8-byte salt and SHA-512 uses a 16-byte salt). The new user SHA-512 hash matches what we expect from the PAM configuration file. The PAM configuration is global and will take effect when users change their passwords. We can force password expirations as a system administrator, but regardless of the method, new and changed passwords on this system will be saved as a more robust SHA-512 hash with salt.

While it seems like we have covered all the particulars of Linux password hashing, we have neglected a whole, very important, file.

For some years, early Linux implementations saved the password hash information in a file called /etc/passwd. Unfortunately, this file could be read by anyone with access to the system, which made it easy for users of the system to access the password hashes of other users. In the late 1970s, some of the more sensitive elements of the /etc/passwd file were removed and saved in the /etc/shadow file we have discussed in this chapter while keeping other, less sensitive details about user accounts in the /etc/passwd file. However, we need to discuss the /etc/passwd file for a few reasons:

- It may assist us with password-cracking
- Some embedded Linux systems (IoT systems) may still use the old method of storing password hash details in the /etc/passwd file

Going back to our test system, we can `cat` the `/etc/passwd` file to see the additional details here, as shown in *Figure 6.6*:

```
jameslv@ubuntu:~$ cat /etc/passwd
root:x:0:0:root:/root:/bin/bash
daemon:x:1:1:daemon:/usr/sbin:/usr/sbin/nologin
bin:x:2:2:bin:/bin:/usr/sbin/nologin
sys:x:3:3:sys:/dev:/usr/sbin/nologin
sync:x:4:65534:sync:/bin:/bin/sync
games:x:5:60:games:/usr/games:/usr/sbin/nologin
man:x:6:12:man:/var/cache/man:/usr/sbin/nologin
lp:x:7:7:lp:/var/spool/lpd:/usr/sbin/nologin
mail:x:8:8:mail:/var/mail:/usr/sbin/nologin
news:x:9:9:news:/var/spool/news:/usr/sbin/nologin
uucp:x:10:10:uucp:/var/spool/uucp:/usr/sbin/nologin
proxy:x:13:13:proxy:/bin:/usr/sbin/nologin
www-data:x:33:33:www-data:/var/www:/usr/sbin/nologin
backup:x:34:34:backup:/var/backups:/usr/sbin/nologin
list:x:38:38:Mailing List Manager:/var/list:/usr/sbin/nologin
irc:x:39:39:ircd:/var/run/ircd:/usr/sbin/nologin
gnats:x:41:41:Gnats Bug-Reporting System (admin):/var/lib/gnats:/usr/sbin/nologin
nobody:x:65534:65534:nobody:/nonexistent:/usr/sbin/nologin
systemd-network:x:100:102:systemd Network Management,,,:/run/systemd:/usr/sbin/nologin
systemd-resolve:x:101:103:systemd Resolver,,,:/run/systemd:/usr/sbin/nologin
systemd-timesync:x:102:104:systemd Time Synchronization,,,:/run/systemd:/usr/sbin/nologin
messagebus:x:103:106::/nonexistent:/usr/sbin/nologin
syslog:x:104:110::/home/syslog:/usr/sbin/nologin
_apt:x:105:65534::/nonexistent:/usr/sbin/nologin
tss:x:106:111:TPM software stack,,,:/var/lib/tpm:/bin/false
uuidd:x:107:114::/run/uuidd:/usr/sbin/nologin
tcpdump:x:108:115::/nonexistent:/usr/sbin/nologin
avahi-autoipd:x:109:116:Avahi autoip daemon,,,:/var/lib/avahi-autoipd:/usr/sbin/nologin
usbmux:x:110:46:usbmux daemon,,,:/var/lib/usbmux:/usr/sbin/nologin
rtkit:x:111:117:RealtimeKit,,,:/proc:/usr/sbin/nologin
dnsmasq:x:112:65534:dnsmasq,,,:/var/lib/misc:/usr/sbin/nologin
cups-pk-helper:x:113:120:user for cups-pk-helper service,,,:/home/cups-pk-helper:/usr/sbin/nologin
speech-dispatcher:x:114:29:Speech Dispatcher,,,:/run/speech-dispatcher:/bin/false
avahi:x:115:121:Avahi mDNS daemon,,,:/var/run/avahi-daemon:/usr/sbin/nologin
kernoops:x:116:65534:Kernel Oops Tracking Daemon,,,:/:/usr/sbin/nologin
saned:x:117:123::/var/lib/saned:/usr/sbin/nologin
nm-openvpn:x:118:124:NetworkManager OpenVPN,,,:/var/lib/openvpn/chroot:/usr/sbin/nologin
hplip:x:119:7:HPLIP system user,,,:/run/hplip:/bin/false
whoopsie:x:120:125::/nonexistent:/bin/false
colord:x:121:126:colord colour management daemon,,,:/var/lib/colord:/usr/sbin/nologin
geoclue:x:122:127::/var/lib/geoclue:/usr/sbin/nologin
pulse:x:123:128:PulseAudio daemon,,,:/var/run/pulse:/usr/sbin/nologin
gnome-initial-setup:x:124:65534::/run/gnome-initial-setup/:/bin/false
gdm:x:125:130:Gnome Display Manager:/var/lib/gdm3:/bin/false
sssd:x:126:131:SSSD system user,,,:/var/lib/sss:/usr/sbin/nologin
jameslv:x:1000:1000:James Leyte-Vidal,,,:/home/jameslv:/bin/bash
systemd-coredump:x:999:999:systemd Core Dumper:/:/usr/sbin/nologin
test:x:1001:1001:Test User,,,:/home/test:/bin/bash
```

Figure 6.6 – The output of the cat /etc/passwd command to see the contents of the /etc/passwd file

Similar to the /etc/shadow file, the /etc/passwd file contains several different pieces of information, separated by colons. The elements, in order from left to right, are as follows:

- **Username:** This is the username that is used to log in to the system. Note that this field is present in both the /etc/passwd and /etc/shadow files.

- The second field is where the password hash would be maintained. This is replaced by an x symbol if the password hash is in the /etc/shadow file.

- **User ID:** The third field is the **User ID** or **UID** and is unique to that particular device.

- **Group ID:** The fourth field is the primary group ID associated with the user.

- The fifth field (which starts with Test User for the test account) is a multi-purpose field that typically contains the user's full name, but may also contain information such as the user's office number, office phone, and home phone. This is referred to as the User ID Info field or the **GECOS** field, a reference to an old OS called the **General Electric Comprehensive Operating Supervisor**.

- **Home directory:** The sixth field is the home directory for that particular user.

- **Shell:** The final field is the shell or command interpreter provided to the user when they log in.

The passwd and shadow files represent the core of what we need to perform cracking against Linux password hashes. Now, let's convert these into a format that our cracking tools expect.

Formatting/converting hashes into their expected formats

The amount of formatting you need to do with the passwd and shadow files will depend on the tool you will use for password cracking. If you wish to use John, you can use the *unshadow* utility to combine the password hash back into the /etc/passwd file as it was originally used. The reason for this is that the /etc/passwd file can contain possible password candidates that John will try if it is aware of them. These are some of the candidates that will be attempted in John's *single crack* mode (along with some mangling rules). Without any other options provided, John will then attempt any wordlist provided, then follow that with incremental mode (that is, brute forcing). To unshadow the files with John, run the following command:

```
unshadow /etc/passwd /etc/shadow > combined.txt
```

Note that if you run the preceding command and the system returns an error that it cannot find the command, run the command from the directory where it is located and add the ./ characters before ./unshadow. You can substitute the preceding values with wherever you have located the passwd and shadow files and name the combined file anything you like. A sample output of the unshadow command is shown in the following code:

```
xxxxxx@slingshot:~$ sudo unshadow passwd shadow > combined
xxxxxx@slingshot:~$ cat combined
```

```
root:*:0:0:root:/root:/bin/bash
daemon:*:1:1:daemon:/usr/sbin:/usr/sbin/nologin
bin:*:2:2:bin:/bin:/usr/sbin/nologin
sys:*:3:3:sys:/dev:/usr/sbin/nologin
sync:*:4:65534:sync:/bin:/bin/sync
games:*:5:60:games:/usr/games:/usr/sbin/nologin
man:*:6:12:man:/var/cache/man:/usr/sbin/nologin
lp:*:7:7:lp:/var/spool/lpd:/usr/sbin/nologin
mail:*:8:8:mail:/var/mail:/usr/sbin/nologin
news:*:9:9:news:/var/spool/news:/usr/sbin/nologin
uucp:*:10:10:uucp:/var/spool/uucp:/usr/sbin/nologin
proxy:*:13:13:proxy:/bin:/usr/sbin/nologin
www-data:*:33:33:www-data:/var/www:/usr/sbin/nologin
backup:*:34:34:backup:/var/backups:/usr/sbin/nologin
list:*:38:38:Mailing List Manager:/var/list:/usr/sbin/nologin
irc:*:39:39:ircd:/run/ircd:/usr/sbin/nologin
gnats:*:41:41:Gnats Bug-Reporting System (admin):/var/lib/gnats:/usr/
sbin/nologin
nobody:*:65534:65534:nobody:/nonexistent:/usr/sbin/nologin
_apt:*:100:65534::/nonexistent:/usr/sbin/nologin
systemd-network:*:101:102:systemd Network Management,,,:/run/systemd:/
usr/sbin/nologin
systemd-resolve:*:102:103:systemd Resolver,,,:/run/systemd:/usr/sbin/
nologin
messagebus:*:103:104::/nonexistent:/usr/sbin/nologin
systemd-timesync:*:104:105:systemd Time Synchronization,,,:/run/
systemd:/usr/sbin/nologin
pollinate:*:105:1::/var/cache/pollinate:/bin/false
sshd:*:106:65534::/run/sshd:/usr/sbin/nologin
syslog:*:107:113::/home/syslog:/usr/sbin/nologin
uuidd:*:108:114::/run/uuidd:/usr/sbin/nologin
tcpdump:*:109:115::/nonexistent:/usr/sbin/nologin
tss:*:110:116:TPM software stack,,,:/var/lib/tpm:/bin/false
landscape:*:111:117::/var/lib/landscape:/usr/sbin/nologin
usbmux:*:112:46:usbmux daemon,,,:/var/lib/usbmux:/usr/sbin/nologin
lxd:!:999:100::/var/snap/lxd/common/lxd:/bin/false
rtkit:*:113:120:RealtimeKit,,,:/proc:/usr/sbin/nologin
xxxxxx:$6$rounds=656000$5Dn.
O6jpbrUbQUTc$cSBaNyD3kMYOjdYHlmyFCz.7omJonkaVAb5zR.
TKYMZ6TvAET1dErEDHSyUDXxNHySJuFHbZeCQ/kePu1vgdg1:1001:1001::/home/
xxxxxx:/bin/bash
_flatpak:*:114:121:Flatpak system-wide installation helper,,,:/
nonexistent:/usr/sbin/nologin
dnsmasq:*:115:65534:dnsmasq,,,:/var/lib/misc:/usr/sbin/nologin
lightdm:*:116:125:Light Display Manager:/var/lib/lightdm:/bin/false
```

```
pulse:*:117:127:PulseAudio daemon,,,:/run/pulse:/usr/sbin/nologin
speech-dispatcher:!:118:29:Speech Dispatcher,,,:/run/speech-
dispatcher:/bin/false
gpsd:*:119:20:GPSD system user,,,:/run/gpsd:/bin/false
```

In the preceding code, we can see that the output of the unshadow command shows the elements of the /etc/passwd file, such as the GECOS field, as well as the hashes from the /etc/shadow file.

If you plan to use hashcat, you don't need to use the /etc/passwd file; instead, you can copy the /etc/shadow file's contents and use it for cracking. As noted previously, you will need elevated privileges to access the /etc/shadow file.

We can readily copy and paste a line from the /etc/shadow file to a separate file for cracking. In this test case, we have done this like so:

```
xxxxxx@slingshot:~$ cat hash.txt
xxxxxx:$6$rounds=656000$5Dn.
O6jpbrUbQUTc$cSBaNyD3kMYOjdYHlmyFCz.7omJonkaVAb5zR.
TKYMZ6TvAET1dErEDHSyUDXxNHySJuFHbZeCQ/kePu1vgdg1:1001:1001::/home/
xxxxxx:/bin/bash
```

Here, we can the salt, as well as the hash. We can also see that this particular password is using the SHA512 hash, denoted by 6, as we mentioned earlier in this chapter.

Now, let's finish this chapter by quickly highlighting what you need to start cracking these hashes.

Cracking hashes

For cracking with hashcat, we will need to identify the mode value or type of hash we are trying to crack. In many cases, the easiest way to do this is by running the hashcat --help command from the Terminal. In this case, since we recognize the hash type as SHA512, we can take the output of hashcat -help and grep for the sha512 phrase, as shown in *Figure 6.7*:

```
jameslv@ubuntu:~$ hashcat --help | grep sha512
  1710 | sha512($pass.$salt)                    | Raw Hash, Salted and/or Iterated
  1720 | sha512($salt.$pass)                    | Raw Hash, Salted and/or Iterated
  1730 | sha512(utf16le($pass).$salt)           | Raw Hash, Salted and/or Iterated
  1740 | sha512($salt.utf16le($pass))           | Raw Hash, Salted and/or Iterated
  1800 | sha512crypt $6$, SHA512 (Unix)         | Operating Systems
  6500 | AIX {ssha512}                          | Operating Systems
```

Figure 6.7 – The output of hashcat –help | grep sha512

Here, we can see that the hash type of 1800, noted as 6 for Unix, should provide the right kind of cracking for our sample. If you are working with a different hash type, consult the hashcat documentation for more details, but the basic types you will normally need can be obtained by running the hashcat -help | grep Unix command from the Terminal, as shown in *Figure 6.8*:

```
jameslv@ubuntu:~$ hashcat --help | grep Unix
 1500 | descrypt, DES (Unix), Traditional DES        | Operating Systems
  500 | md5crypt, MD5 (Unix), Cisco-IOS $1$ (MD5)     | Operating Systems
 3200 | bcrypt $2*$, Blowfish (Unix)                  | Operating Systems
 7400 | sha256crypt $5$, SHA256 (Unix)                | Operating Systems
 1800 | sha512crypt $6$, SHA512 (Unix)                | Operating Systems
```

Figure 6.8 – Common Unix/Linux hash types

As we can see from the preceding screenshot, we can leverage the following modes based on the hash types in the /etc/shadow file:

- MD5 (1):mode 500

- Blowfish bcrypt ($2*$):mode 3200

- SHA256 (5):mode 7400

- SHA512 (6):mode 1800

If you are cracking multiple hashes for a given hash type, it may be faster to gather them into one file and crack them all at once.

Taking our previous example of the jameslv user with their SHA512 hash, we can take this hash into a file for cracking with hashcat and crack this against mode 1800. To do this with a provided wordlist, such as rockyou, we can issue the following command, assuming that the hash has been written to a file called hash.txt and we are using the rockyou wordlist locally:

```
hashcat -m 1800 -a 0 hash.txt rockyou.txt
```

It is important to note that cracking speeds will vary greatly, depending on hardware, enabled GPU support, and more. In this case, the output in *Figure 6.9* is from running the single SHA512 hash against rockyou, in a virtual machine, with only the CPU enabled:

```
[s]tatus [p]ause [b]ypass [c]heckpoint [q]uit => s

Session..........: hashcat
Status...........: Running
Hash.Type........: sha512crypt $6$, SHA512 (Unix)
Hash.Target......: $6$dHF.ZUIXY6NrhdYk$38t7KMlEJVqb1m0GJy7gAIxvKlxu5VL...tNVor0
Time.Started.....: Tue May 23 23:12:45 2023 (16 secs)
Time.Estimated...: Wed May 24 04:13:19 2023 (5 hours, 0 mins)
Guess.Base.......: File (rockyou.txt)
Guess.Queue......: 1/1 (100.00%)
Speed.#1.........:      795 H/s (7.43ms) @ Accel:256 Loops:64 Thr:1 Vec:8
Recovered........: 0/1 (0.00%) Digests, 0/1 (0.00%) Salts
Progress.........: 12800/14344384 (0.09%)
Rejected.........: 0/12800 (0.00%)
Restore.Point....: 12800/14344384 (0.09%)
Restore.Sub.#1...: Salt:0 Amplifier:0-1 Iteration:1472-1536
Candidates.#1....: joelmadden -> 120806
```

Figure 6.9 – Hashcat being run against a single SHA512 hash with the rockyou wordlist

In this example, on a virtual machine with no GPU support, the `rockyou` wordlist will be exhausted in about 5 hours – and if the password is not in the wordlist, cracking will not be successful. However, we can leverage rules (discussed in *Chapter 4*) to increase the likelihood of a successful cracking operation. Remember, the more passwords you are trying to crack, the more likely these approaches will be successful for at least some passwords.

Summary

In this chapter, we discussed how to identify, extract, and crack Linux and Unix password hashes. This will allow you to attempt to retrieve the password for a given user account; however, success, as with all cracking operations, is not guaranteed and is still dependent on the wordlist and rules used. In the next chapter, we will discuss a similar tactic with a growing desktop OS – macOS.

7
WPA/WPA2 Wireless Password Cracking

For almost 20 years, wireless communications, most commonly in the form of Wi-Fi, have become an indispensable part of the mobile workforce. While application layer protocols such as **Secure Sockets Layer** (**SSL**) and **Transport Layer Security** (**TLS**) can be used to protect many of these communications, some transmissions over Wi-Fi are unencrypted. We can use protocols such as **Wi-Fi Protected Access** (**WPA**) and **Wi-Fi Protected Access 2** (**WPA2**) to help protect these communications over the wireless layer only (this encryption is stripped away when the communications are no longer wireless). Unfortunately, the implementation of WPA and WPA2 does leave the possibility of cracking available to us. In this case, we can obtain one of the keys utilized to derive the encryption keys used to protect traffic. If this key can be obtained, the WPA/WPA2 network can be accessed (joined) directly, or communications between other devices on the network can be decrypted.

In this chapter, we're going to cover the following main topics:

- WPA/WPA2 architecture
- Obtaining WPA/WPA2 information to crack
- Methods for cracking WPA/WPA2 passphrases

Importantly, we will be focusing on WPA/WPA2 **pre-shared key** (**PSK**) deployments in this chapter. WPA/WPA2 Enterprise Mode deployments are outside the scope of this book (because authentication to these networks can occur in many different ways).

A note about WEP

While uncommon at this point, some Wi-Fi networks are still encrypted using a protocol known as **Wired Equivalent Privacy**, or **WEP**. WEP utilizes an RC4 cipher to create ciphertext from the plaintext and a per-packet key, which is reversed by the receiving device to yield the plaintext. Unfortunately, the implementation of WEP has several critical flaws that allow us to retrieve the WEP key with minimal effort. In this attack, we do not need to guess the WEP key – we can retrieve it with complete accuracy if we can capture a sufficient amount of packets from that WEP-encrypted network.

While outside the overall scope of this chapter, WEP-encrypted networks still exist and may contain items of interest or importance. Thankfully, there is minimal effort required to crack WEP keys. In fact, you only need two components:

- Hardware supporting monitor mode Wi-Fi capture
- Software to capture traffic in monitor mode
- `aircrack-ng` software for cracking WEP keys

Let's talk about each of these in turn.

First, let's briefly talk about monitor mode. Wi-Fi adapters can operate in one of four modes. **Monitor mode** allows for your Wi-Fi adapter to passively capture traffic from networks you are not associated (*joined*) to. This is critical to get the information we need, especially for networks we cannot currently access.

Obtaining hardware and software that supports monitor mode Wi-Fi capture depends on the OS you are using. In Windows, any hardware device with a native Windows driver supports monitor mode capture – it just needs to be enabled by a software package such as `npcap` from the `nmap` team (`https://github.com/nmap/npcap`). This software, in conjunction with Wireshark (`https://www.wireshark.org/`), allows us to perform monitor mode captures in Windows.

In macOS, built-in Wi-Fi cards support monitor mode capture. You can use the `airport` utility in macOS to perform monitor mode captures on a channel you specify.

Finally, in Linux, monitor mode support can be enabled on most Wi-Fi adapters by using a combination of the `iw` and `ip` utilities or by using the `airmon-ng` script included with the `aircrack-ng` suite of tools. We will need the `aircrack` tools for WEP cracking, so let's download and install those now.

You can use `https://github.com/aircrack-ng/aircrack-ng` to download and install `aircrack-ng` (version numbers move slowly with `aircrack`; at the time of this writing, 1.7 is a solid release with minimal issues).

Once installed, `airmon-ng start <wifi interface name>` will put the Wi-Fi interface into monitor mode for capturing packets. At this point, you can use a utility such as `wireshark` or `tcpdump` to capture packets to a file for analysis.

Once you have captured packets for a network with WEP traffic (100k packets will be sufficient, though less may very well be successful), pass that capture file to `aircrack-ng` using the following syntax:

```
aircrack-ng <packetcapturefile>
```

Due to weaknesses in WEP, you do not need wordlists or extensive GPU hardware to crack a WEP key – with sufficient packets, it will typically be done in under a minute (see *Figure 7.1*):

```
$ aircrack-ng aircrack-data.dump
Read 152633 packets.
Choosing first network as target.
Starting PTW attack with 75811 ivs.


                        Aircrack-ng 1.7


            [00:00:15] Tested 1665641 keys (got 8996 IVs)

   KB    depth    byte(vote)
    0   84/ 85    FE(9728) 15(9472) 18(9472) 1C(9472) 25(9472)   40)
    1   22/ 25    92(11264) 17(11008) 41(11008) 69(11008) 73(11008)
    2   19/  2    F3(11520) 2E(11264) 42(11264) 44(11264) 75(11264)
    3   22/  3    ED(11264) 67(11008) 69(11008) 97(11008) F6(11008)
    4   15/ 16    02(11520) 36(11264) 8A(11264) DB(11264) 07(11008)

              KEY FOUND! [ 88:85:74:27:35 ]
```

Figure 7.1 – aircrack-ng running against a captured WEP sample

In this example, you can see where a WEP key was cracked successfully in approximately 15 seconds in relatively modern hardware, without need for GPU acceleration. This is due to vulnerabilities in the WEP implementation and does not require any special equipment or software to exploit.

While WEP network key cracking is not overly complicated, it is still important to understand how to interact with these kinds of networks, as they will pop up periodically. In fact, based on the statistics from the wardriving aggregation site *wigle.net* (`https://wigle.net`), WEP networks are as commonplace as the more recent WPA standard – as evidenced at `https://wigle.net/enc-large2y.html` (see *Figure 7.2*):

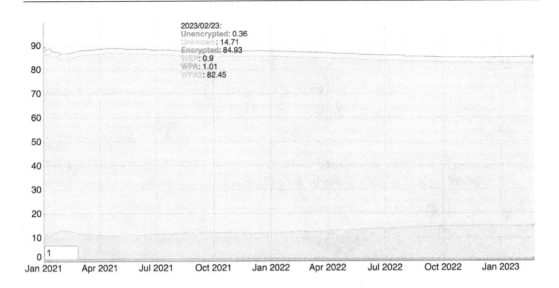

Figure 7.2 – Graph denoting the statistics from wigle.net

As can be seen, WEP networks represent 0.9% of wardrives with little variation over the past 2 years, and WPA with 1.01% and similar variation.

Thankfully, WEP network key cracking is straightforward – collect a sufficient number of packets from the network, and use a tool such as `aircrack-ng` to recover the key. Once the WEP network key is obtained, we can readily use it to join the network or decrypt traffic previously captured.

WPA/WPA2 architecture

In this chapter, we will discuss both WPA and WPA2 in similar terms, mostly because the methods for recovering the network key (also known as a PSK) for these networks are the same for both WPA and WPA2.

> **Important note**
>
> The techniques in this chapter focus on the recovery of the PSK, which is only used for WPA/WPA2 **Personal Mode** networks, not **Enterprise Mode** deployments, which use true authentication, not shared secrets, to authenticate devices/users to the network.

WPA and WPA2 differ in the number of keys and encryption algorithms used to protect data. Most of the remaining processes involved in WPA and WPA2 networks are the same. WPA was designed as a *drop-in* standard to replace WEP, which needed to use the same algorithm as WEP (RC4) due to the computing power of the day. WPA2 was released approximately 5 years later and replaced RC4 with AES for encryption. In both cases, the device is authenticated to the network via knowledge of a PSK that is the same on every device. This PSK can be anywhere from 8 to 63 characters in length.

This PSK is used to derive a **pairwise master key** or **PMK**. As with the PSK, the PMK is the same on every device on this network. The PMK is 256 bits in length and is produced by running the PSK through the PBKDF2 algorithm. This function also takes in as input two additional pieces of data: the SSID or friendly name of the Wi-Fi network, and the length in bytes of that SSID. Consider, for example, a WPA2 personal network with an SSID of `EthicalPasswordCracking`. In this case, the SSID to be input to this function would be "`EthicalPasswordCracking`" (without the quotes), and the length of the SSID would be 23, as the SSID is 23 characters in length. The PSK is run through 4096 rounds of HMAC-SHA1, with the PSK as the key and the SSID and the length of the SSID as additional entropy in the calculation. This process is intentionally computationally expensive, in an attempt to make this process resistant to brute-force and dictionary-based attacks. Unfortunately, this process was designed in the early 2000s, so we will see if this withstands the test of time shortly. Once a PMK is generated, it is typically stored on the endpoint device to avoid the computational overhead of computing it again later.

At this point, we have started with a PSK, which is common on every device in the network. We have now derived a PMK, which, while now a standard length, is still the same on every device – which makes sense – every device has the same PSK and the same SSID, and the same length of that SSID – so it stands to reason that the PMK is also the same on every device. But none of this key material is used to encrypt wireless traffic! We must derive one more key for that purpose – a **pairwise transient key (PTK)**. A PTK takes multiple inputs to the creation of this key:

- The PMK
- The MAC address of the access point
- The MAC address of the client or station
- A nonce (128-bit random data value) from the client
- A nonce (128-bit random data value) from the access point

The PTK calculation is performed between the access point and the client (or station) using a process called a **4-way handshake**. This process ensures that both devices introduce some entropy to the process and produce key material that is unique to this pair of devices (access point and station). *All the messages as part of this process are sent without encryption over the air, as key material has not been negotiated between the two parties yet.* As the name implies, there are four steps to this process to help derive this key material:

1. The access point sends a frame to the station, with the nonce from the access point. After this message is received, the station knows everything to calculate the PTK – the PMK, the MAC addresses, the nonce from the access point, and it knows its own nonce, even though it has not sent it over the air yet. So, the station calculates the PTK and prepares the second step.

2. The station sends a frame to the access point, with the nonce from the station, and a message integrity check calculation on the frame is performed using the PTK. When the access point receives this frame, the access point has everything it needs to calculate the PTK, so it recalculates the message integrity check for frame 2 and validates the message integrity check value matches the one sent by the station. At this time, the access point has validated the access point and the station have the same key material. The access point prepares frame 3 to the station.

3. No new material for key derivation is needed; both devices have the PTK. The last two steps are about validation. The access point calculates a message integrity check value for this frame, and the station recalculates and validates this value, confirming the access point has the same PTK as the station. However, we are still not done – while the access point knows it has the same PTK as the client, and the client knows it has the same PTK as the access point, the access point doesn't know that the client knows that it has the same PTK. So, we need one more step – and we prepare frame 4 from the station to the access point.

4. No new key material is transferred, just a message integrity check value on the frame, calculated using the PTK. Upon receipt by the access point, the message integrity check is recalculated and confirmed, and both devices confirm they have the same key material and are ready to use the PTK to protect data in transit over wireless.

For the sake of illustration, see *Figure 7.3* to observe a 4-way handshake, as visualized on the *Wikipedia 802.11i-2004* page at `https://en.wikipedia.org/wiki/IEEE_802.11i-2004`:

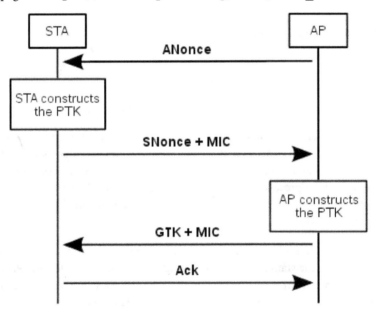

Figure 7.3 – WPA/WPA2 4-way handshake

While the preceding 4-way handshake is the process in the majority of situations, this can be cut short in certain situations in fast-roaming deployments, where a client may quickly move from one access point to the next, using the same SSID, and trying to avoid the 4-way handshake process for the sake of speed. In these cases, some access point deployments may leverage a feature called a **PMK identifier (PMKID)** to avoid the overhead of performing a 4-way handshake every time a new access point in this network is encountered. This process uses the PMK, along with the MAC addresses of the access point and the station with HMAC-SHA-256 to generate a PMKID value. The PMKID is important because we can perform cracking on PMKID-enabled networks without capturing a full 4-way handshake, as we will see in the next section. This process skips the nonces from the access point and station in an attempt to speed up the negotiation process.

Now that we understand the process of WPA/WPA2 authentication, from PSK to PMK to PTK (or PMKID to PTK), let's talk about how we can crack this kind of deployment.

Obtaining WPA/WPA2 information to crack

While this may not be a surprise at this point, we can quickly observe information passed in the clear over the air and use this to perform attacks against the WPA/WPA2 authentication process.

For a traditional authentication process using a 4-way handshake, we can observe the 4-way handshake process and use this to perform cracking of the PSK.

If using the PMKID process, we can capture PMKID information without a valid client and use this to perform cracking of the PSK.

Let's focus on the 4-way handshake process first. Using a Wi-Fi adapter in monitor mode, and set to the correct channel, we can capture traffic, look for a device to connect to the network, and capture the 4-way handshake. During a live capture using a tool such as Wireshark, you can filter on `eapol` as a display filter to see the 4-way handshake during the live capture. Additionally, other tools such as `bettercap` or Kismet can observe the 4-way handshake process and write this out to a packet capture as well – see *Figure 7.4* for the Kismet notation of the 4-way handshake:

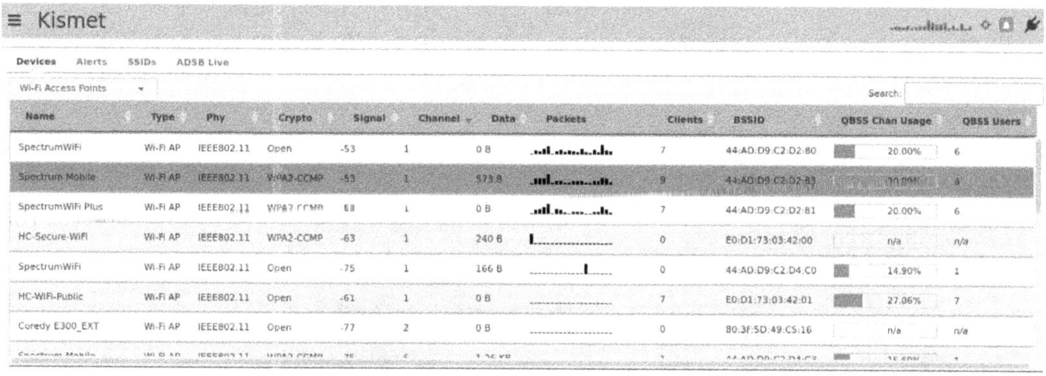

Figure 7.4 – Kismet notation of a captured 4-way handshake by noting the network in red

After noting a 4-way capture, Kismet will write out a copy of a packet capture file with only the 4-way handshake (or PMKID if PMKID was captured). After clicking on the network of interest, you will see an option to download the file – see *Figure 7.5*:

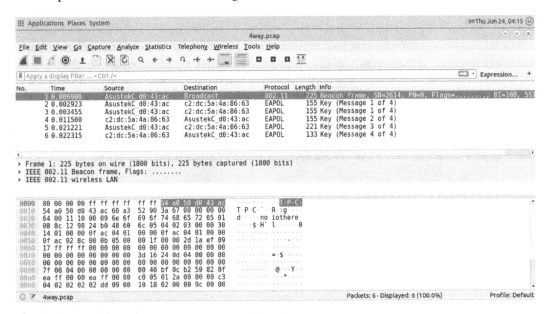

Figure 7.5 – A 4-way handshake captured by Kismet

In this example, we see *step 1* of the 4-way handshake is repeated twice, likely due to transmission issues – however, this does not affect the ability to use this packet capture for cracking. Kismet will also capture the PMKID when supported, as in *Figure 7.6*:

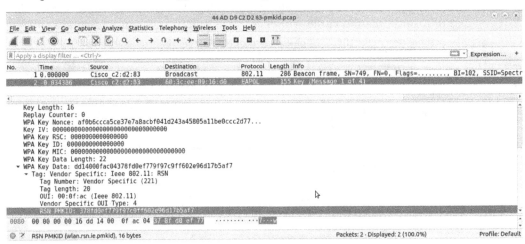

Figure 7.6 – Capture of the PMKID by Kismet

Note that the filename indicating the PMKID capture, as well as the PMKID, is highlighted in the detail on packet 2.

Kismet can be obtained at `https://kismetwireless.net/download/`, and `bettercap` can be obtained at `https://www.bettercap.org/`.

Now that we have our 4-way handshake and/or PMKID, let's move on to cracking.

Methods for cracking WPA/WPA2 passphrases

While John can perform WPA/WPA2 PSK cracking, we will focus in this chapter on cracking these PSKs with hashcat. First, we need a packet capture with either the PMKID or the 4-way handshake. However, we cannot pass the packet capture file directly to hashcat; we need to convert it into a format that hashcat expects. While this capability is not natively included with hashcat, we can use add-on utilities to achieve this objective, such as the terrific `hcxtools` from ZerBea, available at `https://github.com/ZerBea/hcxtools`. Installation is relatively simple; clone the repo followed by `make` and `sudo make install`. Once installed, we will use the `hcxpcapngtool` utility to convert the pcap file, using a syntax like this:

```
hcxpcapngtool -o <output file name> <pcap to convert>
```

In this case, we can call the output file whatever we want – as long as we call it by the same name when we run hashcat in a minute. Now, we can take a look at the output file we created (this output is from a sample file – your output from your network will be different!):

```
WPA*01*2f28a275f277d17904ec948e51012bef*586d8f074e8f*a088b4583fa0*
4d6f62696c6557946692034453846***
WPA*02*4acfe35de7bc8c44b19ba7bfcf2ce152*586d8f074e8f*a088b4583fa0*
4d6f62696c6557946692034453846*
6148801ead3ac326e653a8e5417998245ff5819acd16aee63f0621081325378b*
0103007702010a0000000000000000000288fe22a134055f845914ffa8573f82db
7d34f1dd65a12cae4790738a72c3f8ca0000000000000000000000000000000000
000000000000000000000000000000000000000000000000000000000000000018
30160100000fac040100000fac04010
0000fac023c000000*02
```

You will note in this case we have two lines of output with various elements separated by the * character – one starting with WPA*01* and the next starting with WPA*02*. These two files are two different kinds of hashes – *01* represents PMKID, and *02* is from a traditional 4-way handshake. The next field in PMKID is the PMKID itself (in this case, 2f28a275f277d17904ec948e51012bef); in the 4-way line, this field represents the Message Integrity Check field, which is needed to confirm we have the right PSK. The next two fields are the MAC addresses of the access point and the client, which you will note is the same in both lines (because these two hashes are actually from the same network and client/access point pair). The next two fields are also identical in these examples and are the network name in hex format and the nonce from the access point. At this point, there is no more data in the PMKID line, but the 4-way handshake has some additional data fields. All these fields are documented in the hashcat forums at `https://hashcat.net/forum/thread-10253.html`. One thing to note is that both hashes are written, even if a PMKID is not supported by the network. This means both hashes may not crack, but in this case, we only need one to work.

If you know one of the hash types (PMKID or 4-way) is not present in the `pcap` file, you can filter out the unneeded hash type with another utility from `hcxtools`, `hcxhashtool`.

Now, we can pass this hash file to `hashcat` with the following options:

- `-m` (mode) – This should be set to `22000`
- `-a` (attack mode) – This depends on your approach, as discussed in previous chapters

As with previous cracking types, consider incorporating elements of knowledge about this particular target network, including favorite phrases, sports teams, and so on.

Finally, consider using a **mask attack** (**attack mode 3**) with certain types of networks. Many access points, cellular hotspots, and similar hardware use a somewhat predictable construction format of their PSKs by default. If these defaults are not changed, it may be possible to observe several samples of these devices to determine the pattern and then craft a mask attack based on this. As an example, some cellular hotspots ship with a configured PSK that is 8 hex characters. While a user may change this default, it is worth trying to crack the default configuration, as the 4.2 billion possibilities can be calculated in a very short time on modern GPUs.

Summary

In this chapter, we demonstrated the architecture of WPA and WPA2 PSK networks, as well as how to capture and crack PSKs needed to access these networks or decrypt their traffic. While the computational expense of PSK calculations means that cracking may take longer than other hash types, the computational expense was created in the early 2000s, making cracking significantly easier than it was previously.

In the next chapter, we'll delve into hashes from common web applications like Drupal, WordPress, and Webmin.

8

WordPress, Drupal, and Webmin Password Cracking

Web applications are a cornerstone of modern business and system management. However, their implementation can vary widely, and as employees move companies and change roles, it is possible that some important credentials can be lost and need to be recovered. Thankfully, the related password hashes are often stored in databases or files on the system and can be retrieved. Once you understand the format of that particular application's hashes, you can move on to cracking them. In this chapter, we will highlight three common web applications for system management and content management; however, for research, these principles can be applied to other types of applications as well.

In this chapter, we will cover the following main topics:

- Collecting and formatting WordPress hashes
- Cracking WordPress hashes
- Collecting and formatting Drupal hashes
- Cracking Drupal hashes
- Collecting and formatting Webmin hashes
- Cracking Webmin hashes

Collecting and formatting WordPress hashes

WordPress is a widely used open source **Content Management System** (**CMS**) that allows users to create and manage websites. Its user-friendly interface, extensive plugin ecosystem, and customizable themes make it a versatile platform for bloggers, businesses, and developers. With its robust publishing tools, SEO capabilities, and thriving community, WordPress enables individuals and organizations to build and maintain dynamic websites, from simple blogs to complex e-commerce sites, making it a cornerstone of the web development landscape. Unfortunately, the ease of use of WordPress can often result in misconfigurations regarding WordPress setup when deployed by those less familiar with the technology.

WordPress can be readily installed in a docker container or as a LAMP or WAMP stack (Linux or Windows: **MySQL, Apache**, and **PHP**). In modern WordPress installations, the usernames and passwords are stored locally in the MySQL database used by WordPress. These are saved by default in the wp_users table within MySQL. In order to access the WordPress credentials for the site, a user will need to access the tables in the MySQL database. The easiest way to do this is from the command line using the mysql utility, but this will require elevated privileges in some cases.

After accessing the mysql utility, issue the command show columns from wp_users;. The semicolon (;) character is required at the end of every SQL statement. For more information on using the show columns command, see the MySQL documentation, available at the time of this writing at https://dev.mysql.com/doc/refman/8.0/en/show-columns.html.

After running the show columns from wp_users; command, you will see a list of all the columns in the wp_users table. The table contains information on users, while the columns contain a specific piece of information about all users (for example, a username). Finally, a row in the table represents all the relevant information about one particular entry; in this case, a user with access to the system. The output of the show columns from wp_users; command is shown in *Figure 8.1*:

```
mysql> show columns from wp_users;
+---------------------+-----------------+------+-----+---------------------+----------------+
| Field               | Type            | Null | Key | Default             | Extra          |
+---------------------+-----------------+------+-----+---------------------+----------------+
| ID                  | bigint unsigned | NO   | PRI | NULL                | auto_increment |
| user_login          | varchar(60)     | NO   | MUL |                     |                |
| user_pass           | varchar(255)    | NO   |     |                     |                |
| user_nicename       | varchar(50)     | NO   | MUL |                     |                |
| user_email          | varchar(100)    | NO   | MUL |                     |                |
| user_url            | varchar(100)    | NO   |     |                     |                |
| user_registered     | datetime        | NO   |     | 0000-00-00 00:00:00 |                |
| user_activation_key | varchar(255)    | NO   |     |                     |                |
| user_status         | int             | NO   |     | 0                   |                |
| display_name        | varchar(250)    | NO   |     |                     |                |
+---------------------+-----------------+------+-----+---------------------+----------------+
10 rows in set (0.01 sec)
```

Figure 8.1 – Output of the show columns command against the WordPress users table

Looking more closely at this output, we see that this table contains user information such as their email address, username (in WordPress, this field is called `user_login`), password, display name, and more. However, we are simply here for the username and password. Let's put together another SQL statement to retrieve this information for everyone in the `wp_users` table. For this query, we do not need a lot of extraneous information, so we can ask specifically for the `user_login` and `user_pass` fields from the whole table by using a `select` statement such as `select user_login,user_pass FROM wp_users;`.

This query will retrieve the contents of those two fields for the whole table and display them to the terminal. In this case, we only have one user in this particular installation, so we only have one row returned from our query, as shown in *Figure 8.2*:

```
mysql> select user_login,user_pass FROM wp_users;
+------------+-----------------------------+
| user_login | user_pass                   |
+------------+-----------------------------+
|            | $P$B7sw2SzucH9              |
+------------+-----------------------------+
1 row in set (0.00 sec)
```

Figure 8.2 – Result of the "select user_login,user_pass FROM wp_users;" query

While some of the preceding elements are redacted—sure enough, the `user_pass` fields content looks like a password hash! In fact, it looks like some of the Unix-style hashes we have seen before. However, in WordPress, the password hash of the users defaults to a hashing mechanism known as **phpass**. phpass has been around since the mid-2000s and supports several hashing algorithms; however, out of the box, the WordPress installation typically uses MD5 for hashing and includes a salt as well. The `$P` in the hash specifies phpass, and the `$B` usually indicates the older MD5 version, also known as portable PHP password hashes.

Cracking WordPress hashes

We will only need to copy and paste the hash to crack it using **hashcat**. We can pass the hash directly to hashcat at the command line, encased with single quotes. But first, let's look up the hashcat mode we need to run for this type of hash. Remember, we can run `hashcat -help` and examine the output for this, as shown in *Figure 8.3*:

```
         :~/Desktop$ hashcat --help | grep phpass
  400 | phpass                                        | Generic KDF
```

Figure 8.3 –The "hashcat –help" option, piped to grep to search for "phpass"

With the help of `grep`, we see that mode 400 is required here. So, we will launch hashcat with attack mode 0 (`-a 0`) for a wordlist, mode 400 (`-m 400`) for phpass cracking, our hash, and the wordlist. You can add rules if you need them to increase the likelihood of a successful crack. Without the rules, our command will be `hashcat -a 0 -m 400 'hashvaluehere' rockyou.txt`, as shown in *Figure 8.4*:

```
              :~/Desktop$ hashcat -a 0 -m 400 '$P$B7sw2SzucH9i                ' rockyou.txt
hashcat (v6.2.5) starting

OpenCL API (OpenCL 2.0 pocl 1.8  Linux, None+Asserts, RELOC, LLVM 11.1.0, SLEEF, DISTRO, POCL_DEBUG) - Platform #1 [The pocl project]
=====================================================================================================================================
* Device #1: pthread-Intel(R) Core(TM) i7-1068NG7 CPU @ 2.30GHz, 1439/2942 MB (512 MB allocatable), 2MCU

Minimum password length supported by kernel: 0
Maximum password length supported by kernel: 256

Hashes: 1 digests; 1 unique digests, 1 unique salts
Bitmaps: 16 bits, 65536 entries, 0x0000ffff mask, 262144 bytes, 5/13 rotates
Rules: 1

Optimizers applied:
* Zero-Byte
* Single-Hash
* Single-Salt

ATTENTION! Pure (unoptimized) backend kernels selected.
Pure kernels can crack longer passwords, but drastically reduce performance.
If you want to switch to optimized kernels, append -O to your commandline.
See the above message to find out about the exact limits.

Watchdog: Temperature abort trigger set to 90c
```

Figure 8.4 – hashcat mode 400 being run against a phpass hash with a rockyou wordlist

While portable phpass hashes do not use a modern, robust hashing algorithm by default, multiple rounds of MD5 are used, which does increase the overall time for the cracking operation. However, for a password in our wordlist, after 1 million entries, hashcat was able to crack that password in about 10 minutes, as shown in *Figure 8.5*:

```
$P$B7sw2SzucH9                    :

Session..........: hashcat
Status...........: Cracked
Hash.Mode........: 400 (phpass)
Hash.Target......: $P$B7sw2SzucH9
Time.Started.....: Mon Jun 26 04:37:30 2023 (9 mins, 34 secs)
Time.Estimated...: Mon Jun 26 04:47:04 2023 (0 secs)
Kernel.Feature...: Pure Kernel
Guess.Base.......: File (rockyou.txt)
Guess.Queue......: 1/1 (100.00%)
Speed.#1.........:     1835 H/s (8.67ms) @ Accel:64 Loops:1024 Thr:1 Vec:16
Recovered........: 1/1 (100.00%) Digests
Progress.........: 1045632/14344384 (7.29%)
Rejected.........: 0/1045632 (0.00%)
Restore.Point....: 1045504/14344384 (7.29%)
Restore.Sub.#1...: Salt:0 Amplifier:0-1 Iteration:7168-8192
Candidate.Engine.: Device Generator
Candidates.#1....: Nelson05 -> NOVEMBER15
Hardware.Mon.#1..: Util: 97%

Started: Mon Jun 26 04:37:12 2023
Stopped: Mon Jun 26 04:47:06 2023
```

Figure 8.5 – phpass hash cracked using hashcat

In this case, hashcat was trying a word from the wordlist about 1,800 times a second. It is worth noting that this was CPU cracking in a virtual machine, and cracking with a GPU would have been much faster; for example, on the author's Geforce RTX 3080, it took less than one second.

John the Ripper can also crack WordPress hashes by passing the —format=phpass directive, along with the hash, by using the format user:hash in a file and the wordlist for cracking. The syntax for John would be john —format=phpass wordpress.hash --wordlist=rockyou.txt.

See *Figure 8.6* for a visual of the syntax:

```
                 :~/Desktop$ john --format=phpass wordpress.hash --wordlist=rockyou.txt
Using default input encoding: UTF-8
Loaded 1 password hash (phpass [phpass ($P$ or $H$) 512/512 AVX512BW 16x3])
Cost 1 (iteration count) is 8192 for all loaded hashes
Will run 2 OpenMP threads
Press 'q' or Ctrl-C to abort, 'h' for help, almost any other key for status
0g 0:00:00:11 4.27% (ETA: 04:54:37) 0g/s 63825p/s 63825c/s 63825C/s Zaq1xsw2..VACACIONES
                 (            )
1g 0:00:00:16 DONE (2023-06-26 04:50) 0.06116g/s 63953p/s 63953c/s 63953C/s OMEGAS..NOTORIOUS1
Use the "--show --format=phpass" options to display all of the cracked passwords reliably
Session completed.        _
```

Figure 8.6 – phpass hash cracked using John

Now that we have explained how to obtain and crack WordPress hashes, let's talk about another technology commonly used in this space: **Drupal**.

Collecting and formatting Drupal hashes

Drupal is a highly flexible and open source CMS that allows individuals and organizations to create and manage websites and web applications. With its modular architecture and a vast library of contributed modules and themes, Drupal offers customization and scalability options to a business using the site. It can even be used for building complex and dynamic websites and e-commerce platforms. Drupal's strong focus on security, content management, and extensibility, combined with an active developer community, make it a popular choice for those seeking a powerful and adaptable CMS solution for a wide range of web projects.

Like WordPress, Drupal can be deployed in a LAMP, LEMP (substituting nginx for Apache), or WAMP stacks. For our purposes, we are most concerned with the MySQL stack, where the password hashes are stored. We can use the mysql utility to access the database layer to retrieve the password hashes. Starting with the mysql> prompt, we run the show databases; command to see what databases exist in the installation. In this case, we see the Drupal database, which is our target, as shown in *Figure 8.7*:

```
mysql> SHOW DATABASES;
+--------------------+
| Database           |
+--------------------+
| drupal             |
| information_schema |
| mysql              |
| performance_schema |
| sys                |
+--------------------+
5 rows in set (0.00 sec)
```

Figure 8.7 – "SHOW DATABASES" command in our Drupal installation

> **Important note**
>
> As with the WordPress installation, we will need to end all our SQL query statements with a semicolon (;) character to let mysql know this is the end of the command.

We can see the database—Drupal—that we need to interact with. So, let's use the use Drupal command to switch to the Drupal database, as shown in *Figure 8.8*:

```
mysql> USE drupal
Reading table information for completion of table and column names
You can turn off this feature to get a quicker startup with -A

Database changed
```

Figure 8.8 – Switching to the Drupal database

Next, we would use the show tables; command, which will show us all the tables in the Drupal database. Unfortunately for us, there are a lot of tables in the Drupal database. Even looking for tables with user in the name yields several results, as shown in *Figure 8.9*:

Figure 8.9 – Partial listing of tables in the Drupal database

Upon casual observation, one might expect the `users` table to have the password hashes. Surprisingly, this is not the case, and the table we are looking for is `users_field_data`. Let's grab the information about the columns from that table with a `show columns for users_field_data;` statement, as shown in *Figure 8.10*:

```
mysql> show columns from users_field_data;
+-------------------------+-----------------+------+-----+---------+-------+
| Field                   | Type            | Null | Key | Default | Extra |
+-------------------------+-----------------+------+-----+---------+-------+
| uid                     | int unsigned    | NO   | PRI | NULL    |       |
| langcode                | varchar(12)     | NO   | PRI | NULL    |       |
| preferred_langcode      | varchar(12)     | YES  |     | NULL    |       |
| preferred_admin_langcode| varchar(12)     | YES  |     | NULL    |       |
| name                    | varchar(60)     | NO   | MUL | NULL    |       |
| pass                    | varchar(255)    | YES  |     | NULL    |       |
| mail                    | varchar(254)    | YES  | MUL | NULL    |       |
| timezone                | varchar(32)     | YES  |     | NULL    |       |
| status                  | tinyint         | YES  |     | NULL    |       |
| created                 | int             | NO   | MUL | NULL    |       |
| changed                 | int             | YES  |     | NULL    |       |
| access                  | int             | NO   | MUL | NULL    |       |
| login                   | int             | YES  |     | NULL    |       |
| init                    | varchar(254)    | YES  |     | NULL    |       |
| default_langcode        | tinyint         | NO   |     | NULL    |       |
+-------------------------+-----------------+------+-----+---------+-------+
15 rows in set (0.00 sec)
```

Figure 8.10 – Columns from the "users_field_data" table

Finally, we see a field called `pass` with a 255-character-maximum field. This seems like a good candidate, so, as was the case with WordPress, let's do a `select` statement to grab the **uid**, **name**, and **pass** fields, with the query `select uid,name,pass from users_field_data;`, as shown in *Figure 8.11*:

```
mysql> select uid,name,pass from users_field_data;
+-------+----------+----------------------------------------+
| uid   | name     | pass                                   |
+-------+----------+----------------------------------------+
|    0  |          | NULL                                   |
|    1  | jameslv  | $2y$10$JljaHGRP                        |
```

Figure 8.11 – Output from the usernames and passwords in the "users_field_data" table

We now have access to the password hashes, and the format seems somewhat familiar.

Cracking Drupal hashes

Taking a closer look at the format, as well as the Drupal documentation, tells us that this is a **bcrypt** hash, which is denoted with the `$2` sequence. The `10` appears to indicate rounds of work, followed by the hash data. This could be interesting, as bcrypt is an algorithm designed to be computationally expensive for password hashing.

Let's grab our hash, save it in a file, and try to find a compatible format to use for cracking. When looking at the hashcat documentation and examples, while a Drupal 7 mode exists, the hash format is very different from ours. However, running `hashcat -help` and piping the output to `grep bcrypt` yields a few possibilities, as shown in *Figure 8.12*:

```
            slingshot:~/Desktop$ hashcat --help | grep bcrypt
 3200 | bcrypt $2*$, Blowfish (Unix)                | Operating System
25600 | bcrypt(md5($pass)) / bcryptmd5              | Forums, CMS, E-Commerce
25800 | bcrypt(sha1($pass)) / bcryptsha1            | Forums, CMS, E-Commerce
```

Figure 8.12 – bcrypt modes in hashcat

In this case, the closest match seems to be mode 3200, which directly references the $2 format we see in the hash, so this is likely to be a good candidate. Let's try this with our virtual machine hashcat, using the command hashcat -a 0 -m 3200 Drupal.hash rockyou.txt, as shown in *Figure 8.13*:

```
                    :~/Desktop$ hashcat -a 0 -m 3200 drupal.hash rockyou.txt
hashcat (v6.2.5) starting

OpenCL API (OpenCL 2.0 pocl 1.8  Linux, None+Asserts, RELOC, LLVM 11.1.0, SLEEF, DISTRO, POCL_DEBUG) - Platform #1 [The pocl project]
===========================================================================================================================
* Device #1: pthread-Intel(R) Core(TM) i7-1065NG7 CPU @ 2.30GHz, 1439/2942 MB (512 MB allocatable), 2MCU

Minimum password length supported by kernel: 0
Maximum password length supported by kernel: 72

Hashes: 1 digests; 1 unique digests, 1 unique salts
Bitmaps: 16 bits, 65536 entries, 0x0000ffff mask, 262144 bytes, 5/13 rotates
Rules: 1

Optimizers applied:
* Zero-Byte
* Single-Hash
* Single-Salt
```

Figure 8.13 – Running hashcat mode 3200 against our Drupal hash

So, we expect slower performance than with phpass for cracking; let's get our status and check. Once things are up and running, hit the S key to check the status, as shown in *Figure 8.14*:

```
[s]tatus [p]ause [b]ypass [c]heckpoint [f]inish [q]uit => s

Session..........: hashcat
Status...........: Running
Hash.Mode........: 3200 (bcrypt $2*$, Blowfish (Unix))
Hash.Target......: $2a$05$LhayLxezLhK1LhWvKxCyLOj0j1u.Kj0jZ0pEmm134uzr...vQJLF6
Time.Started.....: Wed Jun  5 12:17:27 2024 (1 min, 2 secs)
Time.Estimated...: Wed Jun  5 15:01:40 2024 (2 hours, 43 mins)
Kernel.Feature...: Pure Kernel
Guess.Base.......: File (rockyou.txt)
Guess.Queue......: 1/1 (100.00%)
Speed.#1.........:     1456 H/s (2.22ms) @ Accel:4 Loops:32 Thr:1 Vec:1
Recovered........: 0/1 (0.00%) Digests
Progress.........: 89820/14344390 (0.63%)
Rejected.........: 0/89820 (0.00%)
Restore.Point....: 89820/14344390 (0.63%)
Restore.Sub.#1...: Salt:0 Amplifier:0-1 Iteration:0-32
Candidate.Engine.: Device Generator
Candidates.#1....: baghera -> badbaby
Hardware.Mon.#1..: Util: 76%
```

Figure 8.14 – hashcat performance for mode 3200 in a virtual machine

Cracking is a good bit slower than phpass, which makes sense. This demonstrates the areas where bcrypt shines, by making each guess take longer, resulting in the password taking longer to crack. Since we have access to a GPU, let's see if we can speed things up and run this same hash with mode 3200 against an RTX 3080 GPU, using the same command:

`hashcat -a 0 -m 3200 Drupal.hash rockyou.txt` (output, as shown in *Figure 8.15*):

```
$2y$10$JljaHG                                                          :

Session..........: hashcat
Status...........: Cracked
Hash.Mode........: 3200 (bcrypt $2*$, Blowfish (Unix))
Hash.Target......: $2y$10$JljaHG
Time.Started.....: Fri Sep 15 00:17:56 2023 (6 mins, 34 secs)
Time.Estimated...: Fri Sep 15 00:24:30 2023 (0 secs)
Kernel.Feature...: Pure Kernel
Guess.Base.......: File (.\rockyou.txt)
Guess.Queue......: 1/1 (100.00%)
Speed.#1.........:     2657 H/s (8.86ms) @ Accel:1 Loops:16 Thr:24 Vec:1
Recovered........: 1/1 (100.00%) Digests (total), 1/1 (100.00%) Digests (new)
Progress.........: 1046112/14344384 (7.29%)
Rejected.........: 0/1046112 (0.00%)
Restore.Point....: 1044480/14344384 (7.28%)
Restore.Sub.#1...: Salt:0 Amplifier:0-1 Iteration:1008-1024
Candidate.Engine.: Device Generator
Candidates.#1....: PORTA -> MyLifeSucks
Hardware.Mon.#1..: Temp: 64c Fan: 39% Util: 97% Core:1935MHz Mem:9251MHz Bus:8
```

Figure 8.15 – Our Drupal hash against a 3080 GPU

This is a lot more respectable. As we can see here, we were able to recover the password in about 7 minutes using our GPU, and we went from 1456 guesses a second to over 2600. Again, GPU acceleration is significant, although it is not as pronounced as with some other algorithms (bcrypt is harder to speed up using GPUs compared to other algorithms such as SHA512).

The same cracking operations can be run using john with the –format=bcrypt directive, although the CPU cracking speeds of john are similar to hashcat's in this case.

Now that we have our Drupal hashes, let's close out our chapter by talking about another commonly used web application: **Webmin**.

Collecting and formatting Webmin hashes

Webmin is a web-based system administration tool designed to simplify and streamline the management of Unix-based systems, including Linux servers. It provides a **Graphical User Interface (GUI)** that allows users to configure various system settings, manage user accounts, monitor system performance, and perform essential administrative tasks without needing to rely solely on command-line interfaces. Webmin's modular and extensible architecture, along with a wealth of available modules and plugins, make it a valuable resource for system administrators and IT professionals who need a convenient way to administer and maintain their servers and network infrastructure.

Webmin is installed either into a Docker container or natively on a Unix/Linux system. When installed, the root user on the system has access to administer Webmin by default. In this case, the root user will sign in with their existing credentials. While Webmin supports native Unix authentication, Webmin will, by default, create local accounts (separate from Unix) when users are provisioned via the Webmin interface. The creation of new users in Webmin is carried out by going to **Webmin Configuration**, clicking on the three horizontal lines, sometimes referred to as the hamburger menu, and selecting **Webmin Users**, as shown in *Figure 8.16*:

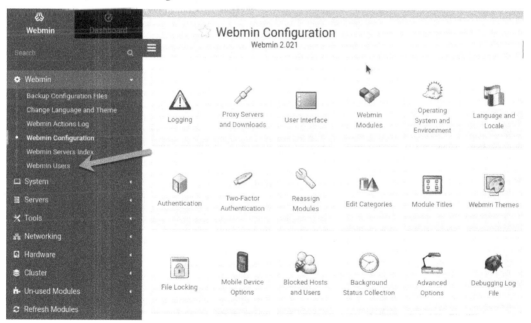

Figure 8.16 – Webmin configuration

Once you're in the Webmin users interface, select **Create a new privileged user**, as shown in *Figure 8.17*:

Figure 8.17 – Creating a new privileged user

From here, you can create a new username, set a password, and force a change on the next login, as shown in *Figure 8.18*:

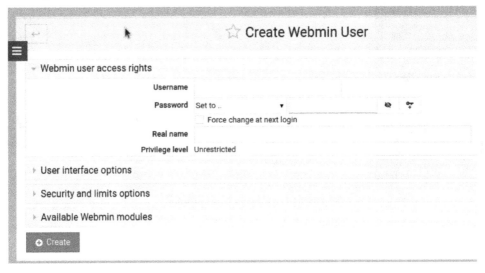

Figure 8.18 – Creating a Webmin user

Once this user is provisioned, the username and password hashes are written to /etc/Webmin/ miniserv.users by default. This file is only accessible to the root user, but it contains the password hashes for the Webmin users on the system. Looking at that file, we see the Webmin users have Unix-style password hashes, with a salt and the indicator for the hash type used, as shown in *Figure 8.19*:

Figure 8.19 – Review of the "miniserv.users" file that contains Webmin user password hashes

Looking at this, we see the $6 value, which indicates that this is a SHA512 hash (as we see in Unix and Linux system password hashing), and the salt (between the second and third $ characters). You will notice the 12 colons (:) at the end of the hash. These are delimiters for the various other fields, similar to what is seen in the Linux /etc/password file. These can be safely removed from the hash when it is moved for cracking.

Cracking Webmin hashes

In order to crack our Webmin hashes, we will leverage the same cracking mode that we would use for a Linux hash of the same type: SHA512. A quick run of hashcat -help and passing the output to grep SHA512 will help us find what we need, as shown in *Figure 8.20*:

Figure 8.20 – Looking for the right mode for UNIX-style sha512 hashes

We see our $6 sequence here under mode 1800, so let's try that for cracking when using our RockYou wordlist. As always, use rules or better-targeted wordlists if you need to! Let's run this hash (saved into a file called Webmin.txt) using the following command:

```
hashcat -a 0 -m 1800 webmin.txt rockyou.txt
```

This is shown in *Figure 8.21*:

```
            :~/Desktop$ hashcat -a 0 -m 1800 webmin.txt rockyou.txt
hashcat (v6.2.5) starting

OpenCL API (OpenCL 2.0 pocl 1.8  Linux, None+Asserts, RELOC, LLVM 11.1.0, SLEEF, DISTRO, POCL_DEBUG) - Platform #1 [The pocl project]
=====================================================================================================================================
* Device #1: pthread-Intel(R) Core(TM) i7-1068NG7 CPU @ 2.30GHz, 1439/2942 MB (512 MB allocatable), 2MCU

Minimum password length supported by kernel: 0
Maximum password length supported by kernel: 256

Hashes: 1 digests; 1 unique digests, 1 unique salts
Bitmaps: 16 bits, 65536 entries, 0x0000ffff mask, 262144 bytes, 5/13 rotates
Rules: 1
```

Figure 8.21 – Starting our Webmin cracking operation against the SHA512 password hash

In this case, hashcat cracked this hash quite quickly, but not for the reasons you might think; the correct hash was quite near the top of the rockyou wordlist, making this deceptively fast. However, it is worth noting that even in GPU cracking operations, SHA512 hashes will crack significantly slower than WordPress phpass hashes, albeit faster than our bcrypt hashes from Drupal. In our virtual machine implementation, our hashing rate is less than half the rate we achieved with the phpass hashes, as you can see in *Figure 8.22*:

```
$6$90213454$RSi346B

Session..........: hashcat
Status...........: Cracked
Hash.Mode........: 1800 (sha512crypt $6$, SHA512 (Unix))
Hash.Target......: $6$90213454$RSi346B664om692pKlrU3.veQiDz7.XVbgtpWfN...22OYD.
Time.Started.....: Thu Sep 14 04:58:47 2023 (0 secs)
Time.Estimated...: Thu Sep 14 04:58:47 2023 (0 secs)
Kernel.Feature...: Pure Kernel
Guess.Base.......: File (rockyou.txt)
Guess.Queue......: 1/1 (100.00%)
Speed.#1.........:      762 H/s (7.31ms) @ Accel:32 Loops:1024 Thr:1 Vec:8
Recovered........: 1/1 (100.00%) Digests
Progress.........: 32/14344384 (0.00%)
Rejected.........: 0/32 (0.00%)
Restore.Point....: 0/14344384 (0.00%)
Restore.Sub.#1...: Salt:0 Amplifier:0-1 Iteration:4096-5000
Candidate.Engine.: Device Generator
Candidates.#1....: 123456 -> butterfly
Hardware.Mon.#1..: Util: 55%
```

Figure 8.22 – Completed cracks of SHA512 hashes; note the slower hashrate compared to phpass!

john should also be able to crack these hashes using the —format=sha512 directive. By considering the previous example, the command would be as follows:

```
john -format=sha512 --wordlist=rockyou.txt webmin.txt
```

At this point, this is mostly a matter of preference, as the cracking speeds of John the Ripper for CPUs will generally be slightly faster, but not extensively.

Summary

In this chapter, we took hashes from several major web-based CMSs and management products and put them through their paces. We saw that different applications could store passwords very differently, and as such, we need to carefully review any documentation, as well as have the appropriate level of access to obtain the hashes from a target system. Once we have them, good wordlists (coupled with good rules) are still vital, as some of these hash types can be very computationally complex, taking a long time to crack. Especially for our stated examples, bcrypt is particularly troublesome and may require the assistance of a GPU.

Furthermore, you have learned how to take these examples and extrapolate them to other applications by understanding where their password hashes are located, extracting them, and determining the right algorithm needed to select the proper cracking mode.

In the next chapter, we will turn to password managers—a great component for keeping passwords safe—and discuss how to try and access these types of storage.

9

Password Vault Cracking

For over 10 years now, an interesting cycle has begun to emerge with regard to the protection of secrets such as passwords. At first, for reasons we have discussed elsewhere in this book, many users would often resort to selecting the same (often weak) password for many sites and services. As data breaches became more and more common, the industry realized that we needed to focus on users selecting unique passwords per site or computer. However, the same password requirements that make a user's password difficult to remember continued to complicate this objective. This is where the password manager comes in.

What if we could offer someone a digital book that could safely store their secrets, and even help them choose good passwords? However, there was still one problem – how did we keep the book safe from others who might want to view those secrets? Password managers secured the book with a key that only the user knew. One passphrase, often known as a vault passphrase, secures access to the digital book or vault, which then offers access to the individual secrets within. The user still needs to remember a password – but just one. Unfortunately, the user sometimes forgets that vault password and doesn't archive it, so we may need to try and retrieve it. Obtaining that information is what we will focus on in this chapter, but we may need to rely on some clues from the user on how the password is constructed to increase our likelihood of success.

In this chapter, we're going to cover the following main topics:

- Collecting KeePass password hashes
- Cracking KeePass password hashes
- Collecting LastPass password hashes
- Cracking LastPass password hashes
- Collecting 1Password password hashes
- Cracking 1Password password hashes

> **Note**
>
> Most modern password managers will prompt you to create a recovery kit when initially creating the password vault. This kit may include printed or electronic copies of the vault passphrase. The purpose of this is to allow for the user's digital possessions to be easily passed on in the case of death or other incapacitation. While this step is optional, a user may have done this, and the vault passphrase may be more readily available than you think. If the user is available to interview, ask whether they ever created such a recovery document, and remind them that it could be with their other sensitive papers. If a little bit of effort upfront might yield the vault passphrase without cracking, that will be an enormous time improvement over even a successful cracking operation. Also, since password managers prompt the user to create a long vault passphrase (20+ characters), cracking a vault passphrase may be difficult if not impossible with current computing technology. Therefore, anything we can do to avoid cracking and recover the vault passphrase via other means is time well spent.
>
> Finally, if we need to proceed to cracking, interview the user if they are available to try and determine any possible characteristics of the passphrase: length, construction, and so forth. Because a vault passphrase needs to be memorable, users will be more likely to use elements they will recall such as sports teams, names of children or pets, and so forth. This may help you construct possible candidates for cracking.

Collecting KeePass password hashes

KeePass is a little different from the other vault types we will discuss, so we will cover it first. KeePass is an open source product (downloads and source are available at `https://sourceforge.net/projects/keepass/`) and its focus and strengths surround its ability to work offline and locally to your machine. This differs from LastPass and 1Password, which use a cloud-based solution for password storage. Contained on your device, KeePass uses a password database file, a single file secured with a vault password or keyfile. The appeal here for some is complete control over your password data (and given some of the recent data breaches surrounding cloud-based password products, this is certainly a worthy perspective to have) by having your KeePass database and dependencies reside locally on your device – and only your device.

However, one weakness of this approach is the idea that KeePass data is not as portable – you can't use a vault on your laptop one moment and your mobile device the next, at least not without configuring some kind of synchronization, which defeats one of the core differences of KeePass. As we will see later in the chapter, even the cloud-based solutions encrypt and decrypt data on your device, for security reasons, so that the vault password never leaves your possession (at least in plaintext). This is why, as we just noted, backups of your vault password are critical for the availability of your vault data.

As a result, KeePass is different from the other solutions based on its approach to the security and ownership of your data – putting the power in the user's hands directly. Your data is less likely to be exposed in a big breach of a cloud provider. However, this comes with risk, meaning data backups, data synchronization, and so forth are also the responsibility of the user, with no one to assist if these actions are not performed. KeePass also lacks some of the neat features of the other options such as form filling and other quality-of-life concerns.

Now that we understand the architecture of KeePass, let's look at the database file from a sample, current KeePass installation – this file was created on Windows, but this should work from any major desktop OS. Moving the file (named Db1.kdbx in this case) to a Linux machine and issuing the file command against the database (file Db1.kdbx) produces an expected result, as shown in *Figure 9.1*:

```
_slingshot:~$ file Db1.kdbx
Db1.kdbx: Keepass password database 2.x KDBX
```

Figure 9.1 – The file command run against a KeePass database

From this output, we can see that KeePass uses a proprietary database structure that the file command recognizes. This file contains password and secret information, as well as the encryption details. However, in this case, we are interested in the encryption data. The location of the KeePass database file is generally set by the user when created, again keeping with the KeePass theme of the user having control over their data. This is not the case with LastPass and 1Password, as we will see later.

As we have discussed several times, the john utility contains many helper scripts to extract the appropriate data and convert these files into output appropriate for cracking. In this case, the keepass2john utility can perform the conversion, though we can use it in both hashcat and John, as we will see shortly. Let us start by running the keepass2john utility against Db1.kdbx, as shown here:

```
******@slingshot:~$ /usr/local/share/john/keepass2john Db1.kdbx >
crackkeepass
```

Now, let's cat the output from crackkeepass and observe the contents:

```
******@slingshot:~$ cat crackkeepass
Db1:$keepass$*2*600000*0*7d4828e7f0576f466cfc0825cf52043b5613fd1f5e
0227edbdca6992c778fb5c*102ca9135997bb36f1b5e2ab0ccd19f695dded55080a
80296576018e42cdf85b*0ce8ea38dc694c6784ea63421e233099*08312fc5bf174
ea3a8ab72374bce90c64e0a6e5c56f6f367cdcc6888cb2cdd8a*54caa49b2077165
74263d8d5a2150667cf4f447b0td70d9714db8a1b0163774b
```

With this, we see that the John conversion utility has extracted the relevant data from the keepass database (Db1.kdbx) for cracking – we have a hash and information that describes the hash type for the cracking programs. With this information, we should be able to move on to cracking.

Cracking KeePass password hashes

Let's try this hash against John and observe the output, by issuing the following command:

```
******@slingshot:~/john crackkeepass --wordlist=rockyou.txt
Warning: detected hash type "KeePass", but the string is also
recognized as "KeePass-opencl"
Use the "--format=KeePass-opencl" option to force loading these as
that type instead
Using default input encoding: UTF-8
Loaded 1 password hash (KeePass [SHA256 AES 32/64])
Cost 1 (iteration count) is 600000 for all loaded hashes
Cost 2 (version) is 2 for all loaded hashes
Cost 3 (algorithm [0=AES 1=TwoFish 2=ChaCha]) is 0 for all loaded
hashes
Will run 4 OpenMP threads
Press 'q' or Ctrl-C to abort, 'h' for help, almost any other key for
status
passwordpasswordpassword (Db1)
1g 0:00:01:05 DONE (2024-04-19 06:34) 0.01523g/s 19.98p/s 19.98c/s
19.98C/s winston..passwordpasswordpassword
Use the "--show" option to display all of the cracked passwords
reliably
Session completed.
```

In this case, the KeePass database was secured with a password of passwordpasswordpassword, which, while not a typically "strong" password, shows the ease of cracking a password of a KeePass vault if the vault passphrase is easily guessable.

The same cracking can be performed in hashcat, while leveraging GPU acceleration if available. Let's start by running hashcat -help | grep KeePass to determine the necessary mode for hashcat, as shown here:

```
******@slingshot:~/hashcat -help | grep KeePass
 13400 | KeePass 1 (AES/Twofish) and KeePass 2 (AES)              |
Password Manager
```

Based on this, it appears that the appropriate mode for hashcat is 13400 for KeePass. Now, let's try the same converted output from keepass2john with the modified rockyou wordlist, as shown here:

```
******@slingshot:~/hashcat -a 0 -m 13400 crackkeepass rockyou.txt
hashcat (v6.2.5) starting

OpenCL API (OpenCL 2.0 pocl 1.8  Linux, None+Asserts, RELOC, LLVM
11.1.0, SLEEF, DISTRO, POCL_DEBUG) - Platform #1 [The pocl project]
====================================================================
====================================================================
```

```
* Device #1: pthread-Intel(R) Core(TM) i7-1068NG7 CPU @ 2.30GHz,
2918/5900 MB (1024 MB allocatable), 4MCU

Minimum password length supported by kernel: 0
Maximum password length supported by kernel: 256

Hashfile 'crackkeepass' on line 1 (Db1:$k...
cf4f447b0fd70d9714db8a1b0163774b): Salt-value exception
No hashes loaded.

Started: Fri Apr 19 07:23:21 2024
Stopped: Fri Apr 19 07:23:21 2024
```

As we can see, this run returned an error. However, the John conversion utilities reference the filename at the start of the converted file, which hashcat does not seem to work well with. Let's remove the filename from the start and rerun hashcat, as shown here:

```
******@slingshot:~/Desktop$ cat crackkeepass
Db1:$keepass$*2*600000*0*7d4828e7f0576f466cfc0825cf52043b5613fd1f5e
0227edbdca6992c778fb5c*102ca9135997bb36f1b5e2ab0ccd19f695dded55080a
80296576018e42cdf85b*0ce8ea38dc694c6784ea63421e233099*08312fc5bf174
ea3a8ab72374bce90c64e0a6e5c56f6f367cdcc6888cb2cdd8a*54caa49b2077165
74263d8d5a2150667cf4f447b0fd70d9714db8a1b0163774b
```

Here's the initial file. Let's remove Db1 from the file and try hashcat again, like this:

```
******@slingshot:~/Desktop$ cat crackkeepass
$keepass$*2*600000*0*7d4828e7f0576f466cfc0825cf52043b5613fd1f5e0227
edbdca6992c778fb5c*102ca9135997bb36f1b5e2ab0ccd19f695dded55080a8029
6576018e42cdf85b*0ce8ea38dc694c6784ea63421e233099*08312fc5bf174ea3a
8ab72374bce90c64e0a6e5c56f6f367cdcc6888cb2cdd8a*54caa49b20771657426
3d8d5a2150667cf4f447b0fd70d9714db8a1b0163774b
```

While it may seem odd that this can prevent a utility from running, remember, computer programs do what we tell them to, and so since hashcat is not expecting the filename at the start of the hash, we need to change the file to meet the expectation of this program. Let's check the output:

```
******@slingshot:~/Desktop$ hashcat -a 0 -m 13400
crackkeepass rockyou.txt
hashcat (v6.2.5) starting

OpenCL API (OpenCL 2.0 pocl 1.8  Linux, None+Asserts, RELOC, LLVM
11.1.0, SLEEF, DISTRO, POCL_DEBUG) - Platform #1 [The pocl project]
============================================================================
============================================================
* Device #1: pthread-Intel(R) Core(TM) i7-1068NG7 CPU @ 2.30GHz,
2918/5900 MB (1024 MB allocatable), 4MCU
```

```
Minimum password length supported by kernel: 0
Maximum password length supported by kernel: 256

Hashes: 1 digests; 1 unique digests, 1 unique salts
Bitmaps: 16 bits, 65536 entries, 0x0000ffff mask,
262144 bytes, 5/13 rotates
Rules: 1

Optimizers applied:
* Zero-Byte
* Single-Hash
* Single-Salt

Watchdog: Temperature abort trigger set to 90c

Host memory required for this attack: 1 MB

Dictionary cache built:
* Filename..: rockyou.txt
* Passwords.: 14344395
* Bytes.....: 139922250
* Keyspace..: 14344388
* Runtime...: 1 sec

$keepass$*2*600000*0*7d4828e7f0576f466cfc0825cf52043b5613fd1f5e0227
edbdca6992c778fb5c*102ca9135997bb36f1b5e2ab0ccd19f695dded55080a8029
6576018e42cdf85b*0ce8ea38dc694c6784ea63421e233099*08312fc5bf174ea3a
8ab72374bce90c64e0a6e5c56f6f367cdcc6888cb2cdd8a*54caa49b20771657426
3d8d5a2150667cf4f447b0fd70d9714db8a1b0163774b:passwordpasswordpassword

Session..........: hashcat
Status...........: Cracked
Hash.Mode........: 13400 (KeePass 1 (AES/Twofish) and KeePass 2 (AES))
Hash.Target......:
$keepass$*2*600000*0*7d4828e7f0576f466cfc0825cf5204...63774b
Time.Started.....: Fri Apr 19 07:29:48 2024 (53 secs)
Time.Estimated...: Fri Apr 19 07:30:41 2024 (0 secs)
Kernel.Feature...: Pure Kernel
Guess.Base.......: File (rockyou.txt)
Guess.Queue......: 1/1 (100.00%)
```

```
Speed.#1.........:          29 H/s (7.60ms) @ Accel:64 Loops:512 Thr:1
Vec:16
Recovered........: 1/1 (100.00%) Digests
Progress.........: 1536/14344388 (0.01%)
Rejected.........: 0/1536 (0.00%)
Restore.Point....: 1280/14344388 (0.01%)
Restore.Sub.#1...: Salt:0 Amplifier:0-1 Iteration:599552-600000
Candidate.Engine.: Device Generator
Candidates.#1....: cuties -> mykids
Hardware.Mon.#1..: Util: 90%


Started: Fri Apr 19 07:29:21 2024
Stopped: Fri Apr 19 07:30:43 2024
```

As we see here, the KeePass vault password can be cracked readily if the password is in previous breach lists or if we can add the appropriate suspicions to the cracking files. However, it is worth noting these speeds are fairly slow and, as such, cracking (especially CPU-based) may take a lot of time. Additionally, when the database is created, the user can choose the encryption algorithm and key derivation function, which can make matters very time-consuming for cracking.

With that, let's move on to one of the big cloud-based providers, LastPass.

Collecting LastPass password hashes

LastPass, while a cloud-hosted service, maintains a local copy of your password vault, which is protected by your vault passphrase. The location for this vault will vary based on the operating system and even browser, but thankfully, LastPass has documented these locations for us on their support site, at this link as of April 2024: https://support.lastpass.com/s/document-item?language=en_US&bundleId=lastpass&topicId=LastPass/FAQ_Data_Storage.html&_LANG=enus. Unlike KeePass, LastPass is a commercial service, which means we need to pay for this service. During the signup process, something very interesting happens – we are prompted to create a vault password that meets specific requirements, including length and complexity, as shown in *Figure 9.2*:

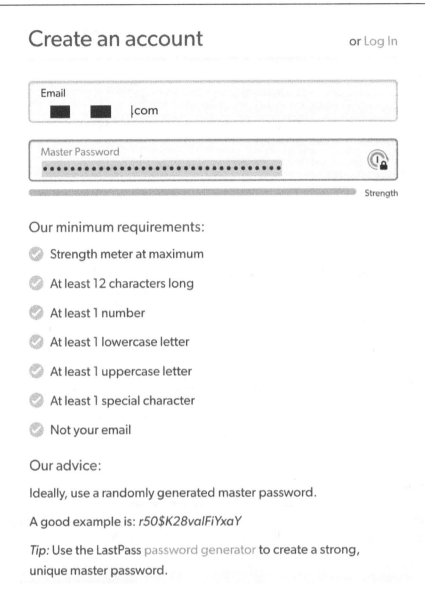

Create an account

or Log In

Email

█████ █████ |com

Master Password

●●

Strength

Our minimum requirements:

- Strength meter at maximum
- At least 12 characters long
- At least 1 number
- At least 1 lowercase letter
- At least 1 uppercase letter
- At least 1 special character
- Not your email

Our advice:

Ideally, use a randomly generated master password.

A good example is: *r50$K28valFiYxaY*

Tip: Use the LastPass password generator to create a strong, unique master password.

Figure 9.2 – LastPass vault password requirements

It is interesting that LastPass has increased these vault password requirements in recent years, especially since guidance from industry sources such as NIST is recommending long passwords over complexity, which is especially relevant to a vault password that the user *has to* remember. Regardless, this is an overall good thing for the user, even though the vault password may be harder to remember. But, if we are trying to recover an existing vault with permission, and the user cannot remember the vault password, this is going to make our work much more difficult.

LastPass continues to iterate on its hash calculation to remain ahead of possible threats. As a result, pulling together the hash information for cracking will take a little bit of effort. Using the browser locations for the preceding vault, we will extract the necessary items for cracking. Many of these techniques are well documented in a blog at `https://markuta.com/cracking-lastpass-vaults/` – many thanks to them for their work in this area. In this example, we are using a vault from a Windows 10 system, using the Google Chrome LastPass extension. This file is located in the following:

```
%LocalAppData%\Google\Chrome\User Data\Default\databases\chrome-
extension_hdokiejnpimakedhajhdlcegeplioahd_0
```

Note that, in this path, `%LocalAppData%` is a defined variable in Windows that will take you to the right path for the logged-in user's local application data store. In this directory is a file usually named `1` – with no file extension. This is a SQLite database and can be accessed from the command line or with a graphical interface tool such as DB Browser for SQLite. Next, we will show you data from the graphical tool, DB Browser for SQLite.

Referencing the great example hashes document from hashcat at `https://hashcat.net/wiki/doku.php?id=example_hashes`, a search shows us that LastPass hashes use mode `6800` and are in the following format: `Hash:rounds:email`.

Here is an example:

```
a2d1f7b7a1862d0d4a52644e72d59df5:500:lp@trash-mail.com
```

The email address is the email address associated with the LastPass account and is required to recover the vault password. As the email address is required, we will need to obtain it via other means, possibly from the user themself. We can find the hash and the number of hashing rounds in the SQLite database. Opening up the database file, we see five tables, as shown in *Figure 9.3*:

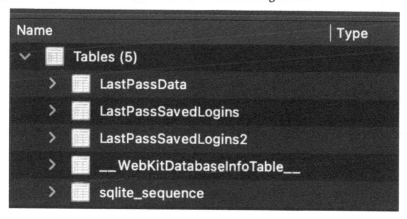

Figure 9.3 – The five tables from the LastPass Google Chrome extension

Drilling into the `LastPassData` table by selecting **Browse Table**, we see several rows of interest to us. First is the line that has `key` listed as the `type` option, as shown in *Figure 9.4*:

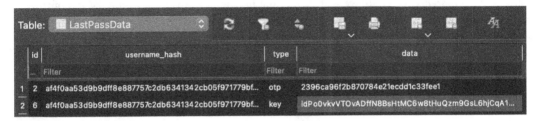

Figure 9.4 – The LastPassData table with the key row highlighted

From here, we need to look at the data in this row. There are two lines of data, and the second line is the one of interest to us, as shown in *Figure 9.5*:

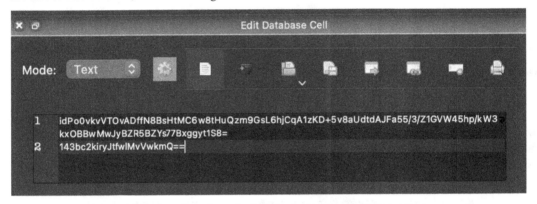

Figure 9.5 – Data from the key row in the LastPassData table

This value (`143bc2kiryJtfwlMvVwkmQ==`) is encoded using Base64. Decoding it and converting it to hexadecimal will yield the hash we need for cracking.

Some operating systems will allow you to do this from the terminal with built-in utilities such as `base64` and `xxd`. In this case, we will do this conversion with CyberChef from the UK GCHQ, available online at `https://gchq.github.io/CyberChef`. Going to this site brings up the CyberChef interface, as shown in *Figure 9.6*:

Figure 9.6 – CyberChef interface

To leverage decoding from Base64 and converting to hex, drag the **From Base64** and **To Hex** widgets from the left column into the **Recipe** column. Then, paste the target value into the **Input** window on the top right. The output should show in the **Output** window on the lower right; if not, click the **BAKE!** button. Once complete, you should see something similar to *Figure 9.7*:

Figure 9.7 – Completed CyberChef decoding and conversion

We now have the hash that we need to input (`d78ddb736922af226d7f094cbd5c2499`) but we still need the rounds and the email address. The rounds can also be obtained in the `LastPassData` table, in the row where the data starts with the word `iterations`. The number to the right of `iterations` = and before the semicolon is the number of hashing rounds used to create the hash – in this case, `600000`, as noted in *Figure 9.8*:

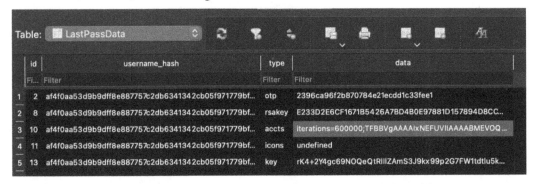

Figure 9.8 – The LastPassData table highlighting iterations (rounds)

With this, we now have all the elements we need to crack this hash and recover the LastPass vault password – in this example, the hash (`d78ddb736922af226d7f094cbd5c2499`), the rounds (`600000`), and the email address, which we will redact for this example. The finalized hash will look like this:

```
d78ddb736922af226d7f094cbd5c2499:600000:email@email.com
```

Finally, we are ready for cracking!

Cracking LastPass hashes

Having saved the preceding hash into a file, we can now proceed to cracking. In hashcat, the mode is `6800`, and we will use an attack mode of `0` with a wordlist in this example, so the command will look like this:

```
hashcat -a 0 -m 6800 lasthash rockyou.txt
```

Here, `lasthash` is a text file containing the hash to crack, and `rockyou.txt` is the wordlist to use. Let's look at the results in *Figure 9.9*:

```
d78ddb736922af226d7f094cbd5c2499:600000:`                @gmail.com:IfYouKnowYouKnow1!one

Session..........: hashcat
Status...........: Cracked
Hash.Mode........: 6800 (LastPass + LastPass sniffed)
Hash.Target......: d78ddb736922af226d7f094cbd5c2499:600000:`      ▬    ...il.com
Time.Started.....: Tue Apr 23 15:05:11 2024 (6 secs)
Time.Estimated...: Tue Apr 23 15:05:17 2024 (0 secs)
Kernel.Feature...: Pure Kernel
Guess.Base.......: File (rockyou.txt)
Guess.Queue......: 1/1 (100.00%)
Speed.#1.........:      162 H/s (5.33ms) @ Accel:128 Loops:1024 Thr:1 Vec:16
Recovered........: 1/1 (100.00%) Digests
Progress.........: 1024/14344389 (0.01%)
Rejected.........: 0/1024 (0.00%)
Restore.Point....: 512/14344389 (0.00%)
Restore.Sub.#1...: Salt:0 Amplifier:0-1 Iteration:599040-599999
Candidate.Engine.: Device Generator
Candidates.#1....: hockey -> abcd1234
Hardware.Mon.#1..: Util: 92%

Started: Tue Apr 23 15:05:09 2024
Stopped: Tue Apr 23 15:05:19_2024
```

Figure 9.9 – Successful cracking of the LastPass hash

You may have noticed that we have used a lot of the John conversion utilities in other chapters to prepare a hash for cracking and wondered why we did not do that here. The reason is that, unfortunately, the `lastpass2john.py` conversion utility has not been updated in many years, and only works on old versions of Firefox and an old version of the LastPass extension.

While we were successful, note the very slow hash speed. While this was CPU-only cracking, it is important to note that attempts against LastPass in general will be slow, so anything the subject can remember about the vault password may be very helpful. Also, bear in mind that newer LastPass accounts are going to have significant complexity requirements, which will make cracking even more difficult. Overall, GPU-based cracking, regardless of the GPU, is going to be the best option for these kinds of hashes.

With that said, let's wrap up this chapter by talking about 1Password.

Collecting 1Password password hashes

1Password is another cloud-based service, similar to LastPass but with some significant architectural differences. 1Password includes a secret key value that is not known to 1Password, which theoretically mitigates the risk of a stolen vault from 1Password directly; however, this secret key is used by the local app or browser extension, which means that this additional protection is really for the cloud infrastructure. We will focus on retrieval from the local vaults in this section.

Creating an account on 1password.com requires you to set the vault password. Unlike LastPass, 1Password only requires a password of 10 characters, with no complexity requirements, as shown in *Figure 9.10*:

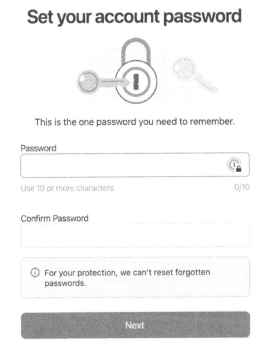

Figure 9.10 – Vault password requirements for 1Password

This means that a password created in 1Password can theoretically be weaker than that of LastPass, given the current construction requirements. 1Password vaults exist to allow the product to work offline and have all the same contents as the cloud-based vault. This vault has changed structure over the years. The early format was referred to as agilekeychain and used multiple files to store data. This was later followed by the cloudkeychain format, which used just one file to store information. Finally, the current structure for the past few versions stores data in a SQLite database. This complicates cracking 1Password vault passwords, as the structure and approach will differ depending on the type of vault you encounter.

In the case of `agilekeychain` and `cloudkeychain`, as well as some early SQLite databases, a conversion utility exists from John called `1password2john.py`. This utility will be installed alongside John with the jumbo patch, as we have referenced in previous chapters. This tool can produce output formats that can be passed to hashcat for faster cracking with modes `6600` (`agilekeychain`) and `8200` (`cloudkeychain`). In both cases, you can review the sample hashes page at `https://hashcat.net/wiki/doku.php?id=example_hashes` for more details. Unfortunately, `1password2john.py` has not been updated to reflect the newer SQLite file structures and will produce an error due to the database schema being different in more modern versions.

Cracking 1Password password hashes

For new structures, we have to get creative. Thankfully, GitHub user *LaurenceGA* developed a tool called `1PasswordWindowsCrack` to extract the appropriate hash elements from a modern 1Password vault in Windows and perform cracking against that. Unfortunately, this approach will not be able to take advantage of GPU-based acceleration from tools such as hashcat, but it will allow us to crack vaults we may know the passwords (or possible passwords) for and access those vaults.

LaurenceGA's code is located at `https://github.com/LaurenceGA/1PasswordWindowsCrack` and will have some prerequisites to install and use the tool, which are defined in the README file. In short, you will need to install Python3 and the C++ compilation tools, and then clone and install the code from *LaurenceGA*'s repo. Once that's done, proceed with the installation of the required Python modules by using `pip install -r requirements.txt`. Once complete, you will need to move the 1Password SQLite database (currently named `1Password10.sqlite`) to the same directory as the cloned repository. You'll also need to supply a word list, which should be named `passwords`, though you can change the name in the script if you prefer.

It's important to note that this utility is attempting passwords against a vault that requires thousands of rounds of hashing, which means that guessing will be very slow. In the following code, we have set up a sample wordlist of 513 passwords (including the known vault password for this particular installation, in the 513th position). The `bruteForcePW.py` utility will open the SQLite database, extract the appropriate value, decode it, split it into the parts that represent the salt, the number of iterations (rounds), and the hash, and use the wordlist to try and reproduce the hash value:

```
Unpacking OPData
Raw opdata:
6f70646174613031400000000000000005be45b0951608333da86c2a722a566a6
**************************************************************
926c187422126284d2edf7ae13d5027e86b9a6e4d60a6846dc9abc4cd8f83ad3a
76a77d6f8e4aaa3e3e430dbb10897d4a3fd39d14bed689ec99bf2bc440a951a54
533003509c4f3b2b0695bca79fb2c6
Header: opdata01
Plain Text length: 64
Initialisation Value: ****************da86c2a722a566a6
Cipher Text ******************************************
******************926c187422126284d2edf7ae13d5027e86b9a
```

```
6e4d60a6846dc9abc4cd8f83ad3a76a77d6f8e4aaa3e3e430dbb10897d4
HMAC digest: a3fd39d14bed689ec99bf2bc440a951********************
OPdata Msg: 6f70646174461303140000000000000005be45b0951608333da86c
2a722a566a6*******************************************************
********926c187422126284d2edf7ae13d5027e86b9a6e4d60a6846dc9abc4cd8
f83ad3a76a77d6f8e4aaa3e3
e430dbb10897d4
HMAC key: c624291706f4bd70f15e9fd3cba0fe6
*******************************
Computed HMAC: 720eede06b4d2c0309153c5a0429c
*********************************
ERROR - Computed HMAC does not match provided value.



Password is: ***********************************



--- 115.6981418132782 seconds ---
```

In this example, 513 password candidates were iterated through in 115 seconds, coming out at under 5 guesses a second. This is positively glacial compared to many of the hash types we have worked with in this book – however, this is not optimal cracking conditions and is not using the GPU. Converting these values to crack with hashcat may yield significantly greater speeds; however, simple tools to do this do not exist at this point.

Summary

In this chapter, we looked at some of the most common password manager (or password vault) utilities. These applications trade off having to memorize many good passwords in exchange for remembering just one good password. However, if this password is not written down or kept safe in other ways, we may be forced to try and retrieve it. Unfortunately, as we have seen, that is easier said than done – between computationally expensive hashing algorithms, flexible numbers of rounds that keep getting increased, and some significant password complexity requirements in some cases, we may have a hard time recovering vault passwords unless the user remembers some elements of its construction, so we can try to leverage techniques such as mask attacks and other methods where partial knowledge can be leveraged. Thankfully, this also translates to greater overall security within these solutions.

In the next chapter, we will look at some common cryptocurrency wallets and evaluate methods to recover their passwords.

10
Cryptocurrency Wallet Passphrase Cracking

In recent years, many of you have probably seen news stories about people who had millions of dollars of some cryptocurrency available to them, usually on a long-forgotten computer. However, they had forgotten the password to the wallet software itself, rendering the cryptocurrency inaccessible. Sometimes users may remember the scheme used to construct the password, which can be helpful as we have seen in discussion of features such as mask attacks; but in any case, we may need to leverage cracking to try and recover access to a cryptocurrency wallet in these situations.

The purpose of this chapter is to give you the tools needed to attempt to recover passphrases for cryptocurrency wallets in some of these situations; while we will not cover every cryptocurrency, you will get a start on some common coins, and from there, you can analyze the given situation at hand and adapt your approach to match.

In this chapter, we will cover the following main topics:

- Cryptocurrencies and blockchain explained
- Collecting and formatting Bitcoin/Litecoin wallet hashes
- Cracking Bitcoin/Litecoin wallet hashes
- Collecting and formatting Ethereum wallet hashes
- Cracking Ethereum wallet hashes

Cryptocurrencies and blockchain explained

Before we begin, we should ground ourselves in a basic understanding of cryptocurrencies and how they work at a high level.

Cryptocurrencies, such as Bitcoin and Ethereum, are digital or virtual forms of money. They use a technology called blockchain to operate, which is like a digital ledger that records all transactions across a network of computers. This technology is what makes cryptocurrencies decentralized, meaning they aren't controlled by one single entity (such as a government or bank), but rather, they are spread out across a network of users worldwide.

To use cryptocurrencies, you need something called a wallet. This isn't like a physical wallet you carry in your pocket; rather, it's software that can be on your computer or smartphone. Your wallet contains your public and private keys. Think of your public key as your email address that you share with others to receive emails, or in this case, cryptocurrency. Your private key is like the password to your email account, but even more secure. You need it to access your cryptocurrency and make transactions. It's crucial to keep your private key safe because if someone else gets it, they can steal your cryptocurrency.

When you want to send or receive cryptocurrencies, transactions are made from one wallet to another. These transactions are then broadcast to the cryptocurrency network and wait to be added to the blockchain. The process of adding transactions to the blockchain involves a method called mining (in many cryptocurrencies). Mining is a process that validates transactions on the blockchain and gives them legitimacy by solving complex mathematical problems; in return, they are often rewarded with newly minted cryptocurrency.

The blockchain is an ordered collection of blocks, where each block contains a number of transactions. Once a block is filled with transactions, it is added to the chain in a linear, chronological order. The blockchain is public, meaning anyone can view it, but it's also secure and tamper-proof. This is because each block contains a unique code called a hash, along with the hash of the previous block. If someone tries to alter a transaction, it will change the block's hash and disrupt the chain's continuity, making the tampering evident.

Cryptocurrencies offer a new way of making transactions that are decentralized, secure, and transparent, thanks to blockchain technology. Wallet software allows users to manage their cryptocurrencies, using private and public keys for security. Transactions are validated through a process called mining, and once verified, they are added to the blockchain.

In this case, the wallet software is our target, attempting to recover the passphrase that unlocks access to the private key.

Collecting and formatting Bitcoin/Litecoin wallet hashes

It is worth noting at the outset that there are many, many types of cryptocurrencies that have come and gone over the years. We are focused on Bitcoin/Litecoin and Ethereum in this chapter for a few reasons.

A large number of cryptocurrencies are based on the codebase of Bitcoin and Litecoin. These are often referred to as 'altcoins' and in some cases are created by using the original Bitcoin/Litecoin code and changing the variable and display names throughout the codebase. Litecoin itself is an altcoin of Bitcoin. However, the similarities in the wallet software of these various coins means the process of extracting hashes and cracking is very similar across coins.

Ethereum is one of the more prominent cryptocurrencies today and has a dedicated cracking mode in hashcat. This means that the knowledge and techniques here may be readily transferrable to other kinds of coins, which is why we cover the ones we do.

If installed locally, the Bitcoin and Litecoin wallets will install their key information in files named `wallet.dat`, located in different places by default.

In Windows 11, the `Bitcoin wallet.dat` will be located at `C:\Users\<Username>\AppData\Roaming\Litecoin\wallets\my wallet\wallet.dat`, as shown in *Figure 10.1*.

Figure 10.1 – Windows 11 Explorer highlighting the location of Bitcoin's wallet.dat file

(In the preceding code, replace `<Username>` with the name of the actual Windows user.)

In Litecoin, the path is similar, and in Windows 11 available by default at `C:\Users\<Username>\AppData\Roaming\Litecoin\wallets\my wallet\wallet.dat`, as seen in *Figure 10.2*:

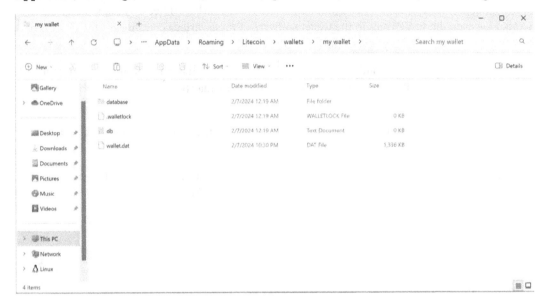

Figure 10.2 – Windows 11 Explorer highlighting the location of Litecoin's wallet.dat file

(In the preceding code, replace `<Username>` with the name of the actual Windows user.)

> **Note**
> While the filenames are the same, they are not interchangeable and represent the wallets for two different cryptocurrencies. Take care, especially if removing them from a target system, to keep them straight as to which is which.

Also worth noting – by default during installation, Bitcoin and Litecoin wallets **do not encrypt** the `wallet.dat` file. If `wallet.dat` is not encrypted, you will be able to move it to another machine and open it directly using the Bitcoin or Litecoin software. However, most guides recommend encrypting the wallet by setting a passphrase, so for the purposes of this chapter, we will show encrypted wallets.

Finally – these techniques target the files that are part of the *official* Bitcoin and Litecoin software, also known as Bitcoin Core and Litecoin Core. Third-party wallet software does exist and is sometimes used, though many of these software packages often store the wallet data off the user's computer, and as a result may be considered less secure by the user. If a third-party wallet is used, these techniques may not work.

Once we have access to the `wallet.dat` file, we can use this file to extract the passphrase hash for cracking. As we have seen in previous chapters, we can use add-on utilities that are included with John to do this. Let's start by moving the `wallet.dat` file to a system where John is installed. Once `wallet.dat` is moved, we can run a quick `file` command against the file to understand what type of file `wallet.dat` is, as shown in *Figure 10.3*:

```
                    :~/Desktop$ file wallet.dat
wallet.dat: Berkeley DB (Btree, version 9, native byte-order)
```

Figure 10.3 – Output from running the file command against wallet.dat

We can see here that the file does have some organization but not something we can natively parse. However, John can parse this and extract the hash with a utility called `bitcoin2john.py`. We can call this from wherever the John scripts are located on the system and pass it the file that we want to extract the hash from (in this case, `wallet.dat`). The command for this is as follows:

`python3 /opt/john/run/bitcoin2john.py wallet.dat`

This is also shown in *Figure 10.4*:

```
                :~/Desktop$ python3 /opt/john/run/bitcoin2john.py wallet.dat
$bitcoin$64$95fc710cea0284fd6c564ee05d9e844733f3933c361666f134fbfc83c7a3f9e9$16$489479e6f520c23d$193278$2$00$2$00
```

Figure 10.4 – bitcoin2john output showing the Litecoin hash

Here we can see the hash output that we can use for cracking as extracted from the `wallet.dat` file.

It is important to note that when running `bitcoin2john` on a system for the first time, you may get an error that says **This script needs bsddb3 installed**. What this means is that the system needs a software package that knows how to handle the Berkeley DB format the `wallet.dat` uses. Fortunately, this is quickly solved in Linux by adding a package via the package manager. In the case of both Ubuntu 22.04 and 20.04, the package is called `python3-bsddb3`, and can be installed with the following command in the terminal:

`sudo apt-get install -y python3-bsddb3`

If you receive this error, install the package, and run the script again.

Once we have the hash, we can move on to cracking. In this case, our hash is the following:

`$bitcoin$64$95fc710cea0284fd6c564ee05d9e844733f3933c361666f134fbfc83c7`
`a3f9e9$16$489479e6f520c23d$193278$2$00$2$00`

With our hash in hand, let us move on to cracking.

Cracking Bitcoin/Litecoin wallet hashes

As we have done in previous chapters, we can take the hash seen previously and save it into a file. We will call our `bitcoin.hash`, as seen in *Figure 10.5*:

```
                    :~/Desktop$ cat bitcoin.hash
$bitcoin$64$95fc710cea0284fd6c564ee05d9e844733f3933c361666f134fbfc83c7a3f9e9$16$489479e6f520c23d$193278$2$00$2$00
```

Figure 10.5 – wallet.dat extracted hash saved to a file called bitcoin.hash

Both John and hashcat support cracking extracted Bitcoin and Litecoin passphrases. Let's start with John. We can call John using a wordlist and the file containing the hash, using the format:

```
john bitcoin.hash --wordlist=rockyou.txt
```

This is shown in *Figure 10.6*:

```
                    :~/Desktop$ john bitcoin.hash --wordlist=rockyou.txt
Warning: detected hash type "Bitcoin", but the string is also recognized as "Bitcoin-opencl"
Use the "--format=Bitcoin-opencl" option to force loading these as that type instead
Using default input encoding: UTF-8
Loaded 1 password hash (Bitcoin, Bitcoin Core [SHA512 AES 512/512 AVX512BW 8x])
Cost 1 (iteration count) is 193278 for all loaded hashes
Will run 2 OpenMP threads
Press 'q' or Ctrl-C to abort, 'h' for help, almost any other key for status
password         (?)
1g 0:00:00:00 DONE (2024-02-08 07:27) 11.11g/s 177.7p/s 177.7c/s 177.7C/s 123456..jessica
Use the "--show" option to display all of the cracked passwords reliably
Session completed.
```

Figure 10.6 – John being run against the bitcoin.hash file

We see here that John has reliably (and quickly) cracked the hash associated with this Bitcoin wallet, and that it was set with a password of `password` – not the best choice but fair enough for demonstration purposes. While John notes in its output that the hash matches a couple of formats, it was cracked successfully without having to explicitly specify a format.

We can do the same kind of cracking with hashcat. Take the extracted hash, and let's check hashcat to find the correct mode to use by taking the help output and `grep` for the word bitcoin with the following command:

```
hashcat -h | grep Bitcoin
```

This shown in *Figure 10.7*:

```
                    :~/Desktop$ hashcat -h | grep Bitcoin
 11300 | Bitcoin/Litecoin wallet.dat                         | Cryptocurrency Wallet
```

Figure 10.7 – hashcat help to identify the mode

We can see that hashcat cracking for Bitcoin and Litecoin (and indeed, many altcoins) will use mode 11300. Let's construct a command to crack that with hashcat, as follows:

```
hashcat -a 0 -m 11300 bitcoin.hash rockyou.txt
```

This is shown in *Figure 10.8*:

```
                    :~/Desktop$ hashcat -a 0 -m 11300 bitcoin.hash rockyou.txt
hashcat (v6.2.5) starting

OpenCL API (OpenCL 2.0 pocl 1.8  Linux, None+Asserts, RELOC, LLVM 11.1.0, SLEEF, DISTRO, POCL_DEBUG) - Platform #1 [The pocl pr
oject]
=========================================================================================================================
======
* Device #1: pthread-Intel(R) Core(TM) i7-1068NG7 CPU @ 2.30GHz, 1439/2942 MB (512 MB allocatable), 2MCU

Minimum password length supported by kernel: 0
Maximum password length supported by kernel: 256

Hashes: 1 digests; 1 unique digests, 1 unique salts
Bitmaps: 16 bits, 65536 entries, 0x0000ffff mask, 262144 bytes, 5/13 rotates
Rules: 1

Optimizers applied.
* Zero-Byte
* Single-Hash
* Single-Salt
* Slow-Hash-SIMD-LOOP
* Uses-64-Bit

Watchdog: Temperature abort trigger set to 90c
```

Figure 10.8 – hashcat initializing against bitcoin.hash

As with John, we get the cracked password fairly quickly, identifying our password of password as shown in *Figure 10.9*:

```
* Filename..: rockyou.txt
* Passwords.: 14344394
* Bytes.....: 139921525
* Keyspace..: 14344387
* Runtime...: 1 sec

$bitcoin$64$95fc710cea0284fd6c564ee05d9e844733f3933c361666f134fbfc83c7a3f9e9$16$489479e6f520c23d$193278$2$00$2$00:password

Session..........: hashcat
Status...........: Cracked
Hash.Mode........: 11300 (Bitcoin/Litecoin wallet.dat)
Hash.Target......: $bitcoin$64$95fc710cea0284fd6c564ee05d9e844733f3933...0$2$00
Time.Started.....: Thu Feb  8 07:57:46 2024 (2 secs)
Time.Estimated...: Thu Feb  8 07:57:48 2024 (0 secs)
Kernel.Feature...: Pure Kernel
Guess.Base.......: File (rockyou.txt)
Guess.Queue......: 1/1 (100.00%)
Speed.#1.........:      161 H/s (6.40ms) @ Accel:256 Loops:1024 Thr:1 Vec:8
Recovered........: 1/1 (100.00%) Digests
Progress.........: 256/14344387 (0.00%)
Rejected.........: 0/256 (0.00%)
Restore.Point....: 0/14344387 (0.00%)
Restore.Sub.#1...: Salt:0 Amplifier:0-1 Iteration:0-1
Candidate.Engine.: Device Generator
Candidates.#1....: 123456 -> freedom
Hardware.Mon.#1..: Util: 75%
```

Figure 10.9 – hashcat cracking bitcoin.hash using the rockyou wordlist

These techniques will work on Bitcoin and Litecoin Core wallets, as well as many altcoin wallets based on these codebases. One important note is that in the past few years, Bitcoin Core has moved away from BerkeleyDB to sqlite3, which breaks the functionality of `bitcoin2john`. At this time, no public methods are available to extract the hashes from this database format.

Try it yourself! Here's the hash for the Bitcoin wallet – the passphrase is `password`:

```
$bitcoin$64$95fc710cea0284fd6c564ee05d9e844733f3933c361666f134fbfc83c
7a3f9e9$16$489479e6f520c23d$193278$2$00$2$00
```

Collecting and formatting Ethereum wallet hashes

Compared to the relatively old Bitcoin and Litecoin algorithms, Ethereum is a relatively new and popular cryptocurrency. It is an open source, blockchain-based platform popular for its versatility and capability to execute smart contracts. Unlike Bitcoin, which was primarily created as a digital currency, Ethereum's main attraction is its smart contract functionality. Smart contracts are a way of enforcing an agreement in code and documenting it on the blockchain. Contract enforcement is automatic and does not require user intervention. This feature has paved the way for **decentralized applications** (**dApps**) to be built on the Ethereum platform, fostering a rich ecosystem of financial services, games, and apps that operate transparently and without a central point of control.

Ethereum also introduced the concept of a platform token, **Ether** (**ETH**), which serves as the fuel for operating the distributed application platform. Ether is used to compensate participant nodes for computations performed, acting as a form of **gas** that drives the execution of smart contracts. This not only incentivizes miners to contribute their computational power to the network but also regulates the network to ensure that the execution of smart contracts is done efficiently.

As a result of this somewhat different approach to cryptocurrency, Ethereum became quite popular, especially since it was able to be readily mined until 2023 using GPUs. Like Bitcoin and Litecoin, holding the Ethereum token requires a wallet, which can come from many sources. One of the more common wallets is **MyEtherWallet** (or **MEW**), which offers mobile and desktop options for wallet setup and storage. As we discussed earlier in this chapter, while many may elect to store key information in a mobile app or in the cloud, some may choose to host their key material locally on their machines for what they feel is better protection. In MyEtherWallet, you can set up a wallet using the prominently advertised mobile apps as shown in *Figure 10.10*:

Figure 10.10 – MEW prominently displaying the mobile options

When scrolling down, other options such as hardware and software wallets are selected, and MEW even notes that software wallets are **not secure**, owing to the users control over the keys and security settings, as shown in *Figure 10.11*:

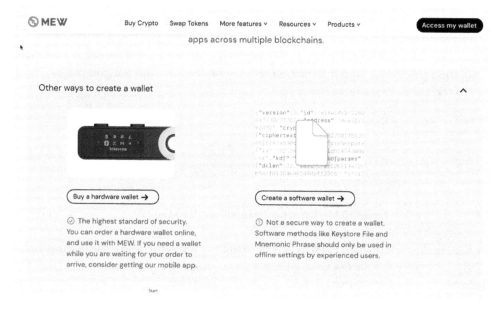

Figure 10.11 – Less prominent MEW options

If selected, the software wallet will prompt the user to create a keystore file, which will contain the hash of the user's passphrase. While MEW labels all this as **not recommended**, some users will choose this option as it gives the user greater control over their wallet. The creation of the passphrase and

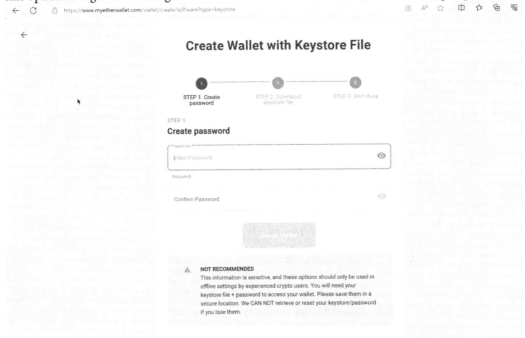

Figure 10.12 – MEW keystore file wallet creation

Once this is complete, MEW will allow you to download your keystore file, which is a **JavaScript Object Notation (JSON)** file with information about your keystore. This file is in plaintext and is readable in a text editor, as shown in *Figure 10.13*:

Figure 10.13 – Opening the keystore in a text editor

Once we have the key information in JSON, we are ready to convert this into a hash format suitable for cracking. Let's move this file (we've saved it as `ether.key`) to a machine running John. Once again, John has an add on utility that can help us convert and extract this hash for cracking; this time the file is called `ethereum2john`. Once the file is moved to the system containing John, you can call the `ethereum2john` utility with the command:

```
python3 /opt/john/run/ethereum2john.py ether.key
```

This is also shown in *Figure 10.14*:

```
                    :~/Desktop$ python3 /opt/john/run/ethereum2john.py ether.key
WARNING: Upon successful password recovery, this hash format may expose your PRIVATE KEY. Do not share extracted hashes with an
y untrusted parties!
ether.key:$ethereum$s*131072*8*1*0e05780bb95d1c1a1c8ad566b0e601cb0f3774bf46de0fa457e10f90f8e04541*b80490cb95f45e8e724182a498292
831f46f69e2ef39e8bf8b1b232a5ad570f7*e80911183ed8fcec3e326758ef86d8748b7279cfeff7b9223ac9cf940a647dcb
```

Figure 10.14 – Using ethereum2john to extract the hash

> **Note**
>
> Replace the path to `ethereum2john` with wherever it is located in your filesystem.

With our hash ready, we can move on to cracking!

Cracking Ethereum wallet hashes

Let's take the output of `ethereum2john` and save it to a file. In this case, we'll call it `ether.hash`, as shown in *Figure 10.15*:

```
                    :~/Desktop$ cat ether.hash
ether.key:$ethereum$s*131072*8*1*0e05780bb95d1c1a1c8ad566b0e601cb0f3774bf46de0fa457e10f90f8e04541*b80490cb95f45e8e724182a498292
831f46f69e2ef39e8bf8b1b232a5ad570f7*e80911183ed8fcec3e326758ef86d8748b7279cfeff7b9223ac9cf940a647dcb
```

Figure 10.15 – Output of ether.hash

Both John and hashcat support cracking this passphrase – let's start with John. As before, we will call John and pass the wordlist and the hash, using the following format:

```
john ether.hash --wordlist=rockyou.txt
```

This is shown in *Figure 10.16*:

```
             :~/Desktop$ john ether.hash --wordlist=rockyou.txt
Using default input encoding: UTF-8
Loaded 1 password hash (ethereum, Ethereum Wallet [PBKDF2-SHA256/scrypt Keccak 512/512 AVX512BW 16x])
Cost 1 (iteration count) is 131072 for all loaded hashes
Cost 2 (kdf [0:PBKDF2-SHA256 1:scrypt 2:PBKDF2-SHA256 presale]) is 1 for all loaded hashes
Will run 2 OpenMP threads
Press 'q' or Ctrl-C to abort, 'h' for help, almost any other key for status
password         (ether.key)
1g 0:00:00:03 DONE (2024-02-08 09:01) 0.2816g/s 9.014p/s 9.014c/s 9.014C/s 123456..butterfly
Use the "--show" option to display all of the cracked passwords reliably
Session completed.
```

Figure 10.16 – Output of John run against ether.hash

Success! We can see the passphrase set for this wallet was `password`. Now let's move on to hashcat. We need to identify the correct mode for cracking with hashcat, so let's look at the hashcat help output, and `grep` for Ethereum:

```
hashcat -h | grep Ethereum
```

This is shown in *Figure 10.17*:

```
                        :~/Desktop$ hashcat -h | grep Ethereum
 16300 | Ethereum Pre-Sale Wallet, PBKDF2-HMAC-SHA256      | Cryptocurrency Wallet
 15600 | Ethereum Wallet, PBKDF2-HMAC-SHA256               | Cryptocurrency Wallet
 15700 | Ethereum Wallet, SCRYPT                           | Cryptocurrency Wallet
```

Figure 10.17 – Output of the `hashcat -h | grep Ethereum` command

Interesting – there is more than one option for Ethereum. Sometimes we will need to make an educated guess, and that is what we will do here. As John's output identified the hash as PBKDF2, we'll try mode 15600 first, with the following command:

```
hashcat -a 0 -m 15600 ether.hash rockyou.txt
```

The output is shown in the following *Figure 10.18*:

```
                -/Desktop$ hashcat -a 0 -m 15600 ether.hash rockyou.txt
hashcat (v6.2.5) starting

OpenCL API (OpenCL 2.0 pocl 1.8  Linux, None+Asserts, RELOC, LLVM 11.1.0, SLEEF, DISTRO, POCL_DEBUG) - Platform #1 [The pocl pr
oject]
=============================================================================================================================
======
* Device #1: pthread-Intel(R) Core(TM) i7-1068NG7 CPU @ 2.30GHz, 1439/2942 MB (512 MB allocatable), 2MCU

Minimum password length supported by kernel: 0
Maximum password length supported by kernel: 256

Hashfile 'ether.hash' on line 1 (ether....8b7279cfeff7b9223ac9cf940a647dcb): Signature unmatched
No hashes loaded.
```

Figure 10.18 – hashcat mode 15600 against ether.hash

We got an error – but if you've done any cracking in the previous chapters, you may know how to fix this. Some of the John conversion utilities add the original filename to the output; in this case, `ether.key`, as shown in *Figure 10.15*. However, John's conversion output doesn't always match what hashcat wants in its formats. Let's check the hashcat example hashes at `https://hashcat.net/wiki/doku.php?id=example_hashes` and look for mode 15600, which is shown in *Figure 10.19*:

```
15000  FileZilla Server >= 0.9.55                  632c4952b8d9adb2c0076c13b57f0c934c80bdc14fc1b4c341c2e0a8fd97c4528729c7bd7ed1268016fc44c3c222445ebb880eca9a6638ea5df74696883a2978.06085161111480S
15100  Juniper/NetBSD sha1crypt                    $sha1$15100$jjDkz0E$E8C7RQAD3NetbSDz7puNAY.5Y2jr
15200  Blockchain, My Wallet, V2                   $blockchain$v2$5000$288$06063152445005516247826076881028813ccf6dcc5793dc0c7a82dcd604c5c3e8d91bea9531e628c2027c56328380c87356f86ae88968f179c366da
15300  DPAPI masterkey file v1 + local context     $DPAPImk$1*1*S-15-21-466364039-425773974-453930460-1925*des3*sha1*24000*b038489dee5ad04e3e3cab4d957258b5*208*cb9b5b7d96a0d2a00305ca403d3fd9c47c
15310  DPAPI masterkey file v1 (context)           $DPAPImk$1*3*S-15-21-407415836-404165111-436049749-1915*des3*sha1*14825*3e86e7d8437c4d5582ff668a83632cb2*208*96ad763b59e67c9f5c3d925e42bbe28a14
15400  ChaCha20 20                                 $chacha20$*0400000000000003*16*0200000000000001*515253545556575$*6b05fe554b0bc3b3
15500  JKS Java Key Store Private Keys (SHA1)       $jksprivk$*SA3AA3C387DD7571727E1725FB09953EF3BEDBD9*0867403720562514024857047678064085141322*81*C3*50DDD9F532430367905C9DE31F81*test
15600  Ethereum Wallet, PBKDF2-HMAC-SHA256         $ethereum$p*262144*323838313731313035343834373738372632343735343738383137303434373S*06eas7ee0a4b9e8abc02c9990e3730827396e8531558ed15bb733faf1
15700  Ethereum Wallet, SCRYPT                     $ethereum$s*262144*1*8*3436383737333838313035343736303637353530323430373235343014363130*8b58d9d15f579faba1cd13dd372faeb51718e7f70735de96f0bcb2c
```

Figure 10.19 – Example hashcat hashes

Note here that the expected hash starts with $ethereum – looking at *Figure 10.15* we see the string is there but there's the filename before it. So, let's edit ether.hash and remove the reference to the ether.key file, so the file looks like *Figure 10.20*:

```
:-/Desktop$ cat ether.hash
$ethereum$s*131072*8*1*0e05780bb95d1c1a1c8ad566b0e601cb0f3774bf46de0fa457e10f90f8e04541*b80490cb95f45e8e724182a498292831f46f69e
2ef39e8bf8b1b232a5ad570f7*e80911183ed8fcec3e326758ef86d8748b7279cfeff7b9223ac9cf940a647dcb
```

Figure 10.20 – Ethereum hash with the filename removed

Now – let's try hashcat again with this modified file, as shown in *Figure 10.21*:

```
                 :-/Desktop$ hashcat -a 0 -m 15600 ether.hash rockyou.txt
hashcat (v6.2.5) starting

OpenCL API (OpenCL 2.0 pocl 1.8  Linux, None+Asserts, RELOC, LLVM 11.1.0, SLEEF, DISTRO, POCL_DEBUG) - Platform #1 [The pocl pr
oject]
===========================================================================================================================
======
* Device #1: pthread-Intel(R) Core(TM) i7-1068NG7 CPU @ 2.30GHz, 1439/2942 MB (512 MB allocatable), 2MCU

Minimum password length supported by kernel: 0
Maximum password length supported by kernel: 256

Hashfile 'ether.hash' on line 1 ($ether...9b7279cfeff7b9223ac9cf940a647dcb): Signature unmatched
No hashes loaded.

Started: Thu Feb  8 09:32:51 2024
Stopped: Thu Feb  8 03.32:51 2824
```

Figure 10.21 – hashcat still doesn't work??

Wait – hashcat still isn't working. Let's take a closer look at that hash in *Figure 10.20*. $ethereum$ is followed by a $s*. Looking at the sample hashes in *Figure 10.19*, mode 15600 should start with an $ethereum$p, but mode 15700 starts with $ethereum$s, so let's try that instead, which is mode 15700, as shown in *Figure 10.22*:

```
          :-/Desktop$ hashcat -a 0 -m 15700 ether.hash rockyou.txt
hashcat (v6.2.5) starting

OpenCL API (OpenCL 2.0 pocl 1.8  Linux, None+Asserts, RELOC, LLVM 11.1.0, SLEEF, DISTRO, POCL_DEBUG) - Platform #1 [The pocl pr
oject]
===========================================================================================================================
======          I
* Device #1: pthread-Intel(R) Core(TM) i7-1068NG7 CPU @ 2.30GHz, 1439/2942 MB (512 MB allocatable), 2MCU

Minimum password length supported by kernel: 0
Maximum password length supported by kernel: 256

Hashes: 1 digests; 1 unique digests, 1 unique salts
Bitmaps: 16 bits, 65536 entries, 0x0000ffff mask, 262144 bytes, 5/13 rotates
Rules: 1

Optimizers applied:
* Zero-Byte
* Single-Hash
* Single-Salt

Watchdog: Temperature abort trigger set to 90c
```

Figure 10.22 – "Look, no errors!"

We see with the new scrypt mode that it seems to load properly – even though John seemed to think the `PBKDF2` mode would work. Does it finish? Let's look at the output in *Figure 10.23*:

```
$ethereum$s*131072*8*1*0e05780bb95d1c1a1c8ad566b0e601cb0f3774bf46de0fa457e10f90f8e04541*b80490cb95f45e8e724182a498292831f46f69e
2ef39e8bf8b1b232a5ad570f7*e80911183ed8fcec3e326758ef86d8748b7279cfeff7b9223ac9cf940a647dcb:password

Session..........: hashcat
Status...........: Cracked
Hash.Mode........: 15700 (Ethereum Wallet, SCRYPT)
Hash.Target......: $ethereum$s*131072*8*1*0e05780bb95d1c1a1c8ad566b0e6...647dcb
Time.Started.....: Thu Feb  8 09:33:49 2024 (2 secs)
Time.Estimated...: Thu Feb  8 09:3 :51 2024 (0 secs)
Kernel.Feature...: Pure Kernel
Guess.Base.......: File (rockyou.txt)
Guess.Queue......: 1/1 (100.00%)
Speed.#1.........:        2 H/s (0.93ms) @ Accel:2 Loops:1024 Thr:1 Vec:1
Recovered........: 1/1 (100.00%) Digests
Progress.........: 4/14344387 (0.00%)
Rejected.........: 0/4 (0.00%)
Restore.Point....: 2/14344387 (0.00%)
Restore.Sub.#1...: Salt:0 Amplifier:0-1 Iteration:130048-131072
Candidate.Engine.: Device Generator
Candidates.#1....: 123456789 -> password
Hardware.Mon.#1..: Util: 81%

Started: Thu Feb  8 09:33:32 2024
Stopped: Thu Feb  8 09:33:53 2024
```

Figure 10.23 – ether.hash cracked in hashcat

We can see the Ethereum wallet has now successfully been cracked by hashcat as well.

Try it yourself! Here's the hash for the Ethereum wallet, for which the plaintext passphrase is `password`:

```
$ethereum$s*131072*8*1*0e05780bb95d1c1a1c8ad566b0e601cb0f3774bf46de0
fa457e10f90f8e04541*b80490cb95f45e8e724182a498292831f46f69e2ef39e8bf
8b1b232a5ad570f7*e80911183ed8fcec3e326758ef86d8748b7279cfeff7b9223ac
9cf940a647dcb
```

Summary

We saw – especially in the Ethereum section of this chapter – a number of missteps: wrong modes, file formats with too much information, and more. So, what can we learn from this series of mistakes?

- Sometimes, tool output can be deceiving
- We may need to slightly alter or tweak tool output to achieve our desired results
- We should look beyond error messages and see how they can help us resolve the problem!

In all the errors we had in this chapter, the information to solve them was right in front of us, we just needed to engage in a little 'trial and error' to achieve the right results. Our tools and conversion utilities are great assets to our work, but we can't just blindly accept their output. We need to look at it and figure out what makes sense in our situation.

In our next and final chapter, let's wrap up everything we've discussed by revisiting our password principles from earlier chapters, with the added insight of our cracking chapters, and provide some recommendations.

Part 3:
Conclusion

In this part, we will conclude by discussing how to produce more defensible passwords, as well as how to build defenses for our password choices.

This part has the following chapter:

11

Protections against Password Cracking Attacks

Depending on how you approached this book, you may have various goals for this chapter. If you are a tester, or someone auditing the strength of passwords in your environment, then you may be concerned about what guidance to provide to the targets of your cracking operations to attempt to make their passwords more resistant to such attacks in the future. If you are trying to recover the password to your long-forgotten Bitcoin wallet, you may be thankful that you did not choose a more robust password at the time!

However, in many scenarios, the task of guiding our users will fall to us. The topic of *how to build a strong password* is not an easy one and the recommendations have changed much over the years. Additionally, our guidance may vary slightly depending on factors such as the robustness of our hashing algorithm and how slow cracking may work against such a construct.

In this chapter, we are going to cover the following topics:

- How to choose a password more resistant to cracking attacks
- Additional protections against cracking attacks

How to choose a password more resistant to cracking attacks

You may note the careful choice of words, using terms such as 'more resistant' in the section title. The reason for this is that there are no absolutes in the security of a computer system. We can make a password more resistant or less resistant to cracking attacks with our choices, but we cannot completely protect against cracking attacks. It is important to set expectations on system security and the possibility of system security, as actions can be taken to reduce – but not eliminate – the risk of a threat to a system.

Given that, how do we make a password more resistant? The answers have been in front of us for years. In the 1980s, the **United States Department of Defense (US DoD)** published a set of *rainbow books* – they covered the management of computer systems, and how to build, audit, and evaluate them. While these books, maintained by the United States **National Institute for Standards and Technology (NIST)**, are no longer considered a standard, they still contain solid guidance for some aspects of system security, and are available online at `https://csrc.nist.gov/pubs/other/1985/12/26/dod-rainbow-series/final`. They were referred to as the rainbow books because each one was printed in a different color cover, making the books color-coded by topic. Password management guidelines were covered in the green book, also identified by its less colorful name, CSC-STD-002-85. This document is now available at `https://csrc.nist.gov/files/pubs/other/1985/12/26/dod-rainbow-series/final/docs/std002.txt`.

Many of you are familiar with the concepts of passwords at your place of employment, and those passwords may have, or have had, requirements such as complexity (number of certain character types), length (how many characters the password can be), and rotation interval (how long you can keep a password before you must change it, and the history of how many previous passwords must exist before a password can be reused). The green book is considered one of the seminal documents on passwords, and yet, if you review it completely, you will note that the document does not contain any references to how often a password should be changed, or how complex the password should be. One of the primary threats at the time revolved around the possibility of a retrieved password hash (or encrypted password) being brute-forced. Brute-forcing represents the most time-consuming type of attack in most cases, so the presumption is, if an attacker can brute-force a password before you are forced to change it, that is an unacceptable risk. As such, passwords should be changed before they can reasonably be brute-forced. As we have seen in the past 10 chapters, though, how long it takes to guess a potential password candidate – whether by brute force or using a wordlist – will vary depending on how many guesses you can make per second. This is a factor of the hashing/encryption algorithm, as well as the speed of the machine doing the guessing. Rather than set different password requirements for different systems, most organizations will set one password policy across all systems.

The green book recommended a password life of up to one year but noted that it should vary depending on many factors. Most organizations coalesced around a standard of relatively short passwords (6 to 8 characters), enforcing basic complexity (such as the use of upper-case characters, lower-case characters, and numbers), and requiring users to change passwords every 90 days. However, this standard did not come from any clear written guidance in the green book.

When I first accessed early networked computer systems in the early 1990s, the 90-day/short-but-complex password requirements were already in place in many systems. The justification was largely this brute-force threat, and as such this guidance became reflected in many places, including in NIST's password standards in older versions of a special publication numbered 800-53.

However, these policies (90-day rotation and complex but relatively short passwords) encourage a few specific risky behaviors from computer end users. The first is leveraging passwords that are easy for users to remember. Let's use a sample organization that leverages 90-day password rotation and password complexity requirements of at least one uppercase letter, one lowercase letter, a number, and 8-character passwords. These complexity requirements may result in a password that is hard for our user to remember. As a result, our user may opt to choose a password that is easier to remember within the constraints required. They could write down their password and stick it on their monitor – but what if they frequently work outside the office? A more portable solution may be needed. The user looks out the window. It's summer where they are. And suddenly, they realize that 'Summer' (with a capital 'S') meets two of the complexity requirements. Now, they just need to get the password to 8 characters, and they need numbers as well in the password. Could they add a 2-digit representation of the year? And would adding that onto the end of the password create a password – 'Summer23' – that matches complexity requirements? It does! And they can even easily rotate it next season, changing the season name to 'Autumn' and keeping the '23'!

Does this password meet the complexity and other password requirements of the organization? Yes, it does. Is this password easy for the user to remember? Yes, it is. Is it a 'good' password? No, it is not. It is readily predictable and will be in the first few lines of an attacker's wordlist. Additionally, common, predictable substitutions in passwords do not increase the complexity of the password, as we noted in *Chapter 1*.

The second risky user behavior that started to increase was the idea of password reuse. Rather than create a difficult-to-remember password on every site they used, many users opted to create one password and use it across multiple systems. Unfortunately, different systems handle security differently, and some do not handle security well at all. This results in breach data that can be combed through for passwords or password hashes using the techniques we described in *Chapter 2*.

In the 2010s, we started to see monumental increases in the capability to crack passwords, largely due to the increased leveraging of more common **Graphics Processor Units** (**GPUs**), which can perform specific types of math very, very quickly. Password hashes were suddenly significantly easier to crack (meaning, more hashes processed in less time) than they used to be. As a result, the security community realized a change was in order. My colleague at the SANS Institute, Lance Spitzner, discusses this in his blog from 2019 at `https://www.sans.org/blog/time-for-password-expiration-to-die/`.

So, how DO we protect against password-cracking attacks? We do so by changing the rules we have operated under for many years. In 2017 (and later revised in 2020), NIST published an updated version of their 800-53B document for digital identity guidelines. This is readily available at `https://pages.nist.gov/800-63-3/sp800-63b.html` as of this writing.

The newest versions of 800-63 are important because they focus on one factor of password security – the length of the password. NIST notes in Appendix A that "Users respond in predictable ways to the requirements imposed by composition rules," as seen in *Figure 11.1*:

A.3 Complexity

As noted above, composition rules are commonly used in an attempt to increase the difficulty of guessing user-chosen passwords. Research has shown, however, that users respond in very predictable ways to the requirements imposed by composition rules [Policies]. For example, a user that might have chosen "password" as their password would be relatively likely to choose "Password1" if required to include an uppercase letter and a number, or "Password1!" if a symbol is also required.

Figure 11.1 – Password complexity assigned in a predictable manner

What does this tell us? *Evidence shows us, time and again, that complexity is largely not relevant and password length produces more cracking-resistant passwords than other methods, while allowing users to more readily remember their passwords.*

The current NIST documentation also suggests that passwords do not need to be rotated frequently if longer passwords are chosen. In fact, the current recommendations largely focus on waiting for evidence of a breached password before rotation of credentials. This is detailed in the current 800-63B document in Section 5.1.1.2 as noted in *Figure 11.2*:

Verifiers SHOULD NOT impose other composition rules (e.g., requiring mixtures of different character types or prohibiting consecutively repeated characters) for memorized secrets. Verifiers SHOULD NOT require memorized secrets to be changed arbitrarily (e.g., periodically). However, verifiers SHALL force a change if there is evidence of compromise of the authenticator.

Figure 11.2 – NIST requirements on password rotation

This recommendation has caused more consternation in the security community, since evidence of breaches does not always surface before credentials are used in attempted compromise. However, one could balance this risk by forcing the rotation of credentials periodically (such as annually) without evidence of compromise, reducing the impact on the user.

Hopefully, the previous pages have shown we need to change our approach to passwords, selecting long but unique passwords that are unique to that particular system. But we still have the problem of how we determine passwords that our users can remember.

In *Chapter 9*, we cracked passwords for password managers, a type of product that stores various secrets for a user. While we saw techniques to crack these 'master passwords', password managers still offer a solid solution to protect the credentials of large amounts of users. These utilities allow us to create unique passwords for every system a user visits. But they do require one password or passphrase to unlock the entire vault of password secrets. In many cases, this still improves the overall security of a given user – rather than craft and remember strong passwords for every system, they only need to craft and remember one master password for their password manager, and use that to access all the other systems they need. This also allows for protecting an individual in the event of their untimely death, as they can share the master password with their family or partners so they can manage their affairs after death. While this feels like 'writing passwords down' – and it is – the password manager implementations protect those credentials in a more robust way than simply writing them down – though truth be told, I would rather a user write their multiple passwords for multiple systems down somewhere than use one password for all systems, which ties their overall security to the weakest link (whatever system has their credentials that is most easy to compromise). A further complication for the user in this scenario is that when a password is compromised, it is difficult for the user to remember any and all other sites where that same credential was reused.

It should be noted that all the defensive techniques we employ to create strong passwords can be defeated if someone can readily access the passwords. We need to ensure overall system security, that passwords cannot be accessed in clear text or readily decrypted by an individual such as a database administrator or system administrator who has access to all (or most) of the data on the system.

Additional protections against cracking attacks

The cracking attacks we have discussed in this book have almost always focused on 'offline attacks' – the idea of cracking an individual's credentials after accessing it from a database or other system. While these attacks are significant, we can attempt to protect systems from unauthorized access to these systems, which somewhat limits this risk. While a worthy objective, the means of securing information systems vary greatly across systems and are outside the scope of this book.

However, even if a user has chosen a poor password and placed that password in a system that was compromised, and their password was cracked, the attacker will still need to access that system to do anything of value. This is where we can employ systems such as **multi-factor authentication** (**MFA**) to protect the user, even if their password is compromised.

You are already familiar with how this works – after authenticating using a username and password, the system will prompt you to finish authenticating with something you have (a token or SMS message) in addition to something you know (a password and user ID). This means that an attacker who has retrieved a password via cracking will not be able to access the target system without also being able to access the MFA token or receive a text message on behalf of the user.

Additionally, hardware and software solutions that build on the FIDO2 standard for passwordless authentication may be appropriate for certain groups of users, with the caveat that losing or misplacing these tokens can complicate logins more significantly than losing an old MFA token.

Additionally, if you assume an attacker will access your systems, you should build monitoring capabilities into the network and systems around that assumption, and look for behavior from user accounts that is out of the norm for that account.

Finally, if an attacker attempts to access a system and they have not successfully cracked a password, look for repeated login attempts from the same user, logins from unusual regions for that account, and anything else that can indicate potential attempts at account takeover.

Summary

In this book, we have introduced you to the concepts of passwords, how they can and should be stored, and various common implementations of passwords so you can recover them if you need to. You have learned how to install cracking tools, as well as when to avoid leveraging cracking tools by using techniques such as OSINT to recover passwords instead. You have also learned how to approach some common use cases of password storage, how to access password hashes, and how to start recovering those passwords. As we have seen though, our success will be dependent on how the original password was constructed, as well as how it was stored. These skills were discussed to allow you to try and retrieve passwords when necessary, to either allow for recovery of important data or information targeted in a penetration-testing/red-team scenario.

However, it is important that both users and systems protect their passwords appropriately to limit the possibility of compromise by unauthorized parties. These techniques that allow us to recover passwords equally allow those without permission to do the same, and the guidelines we've discussed can help limit the possibility of these attacks being successful.

It is critical to remember that these skills should be used with caution, and only with permission! Performing the actions in the previous chapters against a system you are not authorized to test can lead to criminal penalties in many countries. The title of this book is Ethical Password Cracking for a reason – ensure proper, written approval is obtained before attempting to recover passwords from systems.

Index

Symbols

packtpub.com

Subscribe to our online digital library for full access to over 7,000 books and videos, as well as industry leading tools to help you plan your personal development and advance your career. For more information, please visit our website.

Why subscribe?

- Spend less time learning and more time coding with practical eBooks and Videos from over 4,000 industry professionals

- Improve your learning with Skill Plans built especially for you

- Get a free eBook or video every month

- Fully searchable for easy access to vital information

- Copy and paste, print, and bookmark content

Did you know that Packt offers eBook versions of every book published, with PDF and ePub files available? You can upgrade to the eBook version at packtpub.com and as a print book customer, you are entitled to a discount on the eBook copy. Get in touch with us at customercare@packtpub.com for more details.

At www.packtpub.com, you can also read a collection of free technical articles, sign up for a range of free newsletters, and receive exclusive discounts and offers on Packt books and eBooks.

Other Books You May Enjoy

If you enjoyed this book, you may be interested in these other books by Packt:

Defending APIs

Colin Domoney

ISBN: 978-1-80461-712-0

- Explore the core elements of APIs and their collaborative role in API development
- Understand the OWASP API Security Top 10, dissecting the root causes of API vulnerabilities
- Obtain insights into high-profile API security breaches with practical examples and in-depth analysis
- Use API attacking techniques adversaries use to attack APIs to enhance your defensive strategies
- Employ shield-right security approaches such as API gateways and firewalls
- Defend against common API vulnerabilities across several frameworks and languages, such as .NET, Python, and Java

Mastering Cloud Security Posture Management (CSPM)

Qamar Nomani

ISBN: 978-1-83763-840-6

- Find out how to deploy and onboard cloud accounts using CSPM tools
- Understand security posture aspects such as the dashboard, asset inventory, and risks
- Explore the Kusto Query Language (KQL) and write threat hunting queries
- Explore security recommendations and operational best practices
- Get to grips with vulnerability, patch, and compliance management, and governance
- Familiarize yourself with security alerts, monitoring, and workload protection best practices
- Manage IaC scan policies and learn how to handle exceptions

Packt is searching for authors like you

If you're interested in becoming an author for Packt, please visit authors.packtpub.com and apply today. We have worked with thousands of developers and tech professionals, just like you, to help them share their insight with the global tech community. You can make a general application, apply for a specific hot topic that we are recruiting an author for, or submit your own idea.

Share Your Thoughts

Now you've finished *Ethical Password Cracking*, we'd love to hear your thoughts! Scan the QR code below to go straight to the Amazon review page for this book and share your feedback or leave a review on the site that you purchased it from.

https://packt.link/r/1804611263

Your review is important to us and the tech community and will help us make sure we're delivering excellent quality content.

Download a free PDF copy of this book

Thanks for purchasing this book!

Do you like to read on the go but are unable to carry your print books everywhere?

Is your e-book purchase not compatible with the device of your choice?

Don't worry!, Now with every Packt book, you get a DRM-free PDF version of that book at no cost.

Read anywhere, any place, on any device. Search, copy, and paste code from your favorite technical books directly into your application.

The perks don't stop there, you can get exclusive access to discounts, newsletters, and great free content in your inbox daily

Follow these simple steps to get the benefits:

1. Scan the QR code or visit the following link:

https://packt.link/free-ebook/9781804611265

2. Submit your proof of purchase.
3. That's it! We'll send your free PDF and other benefits to your email directly.

www.ingramcontent.com/pod-product-compliance
Lightning Source LLC
Chambersburg PA
CBHW080532060326
40690CB00022B/5107